ANIMAL HEMATOTOXICOLOGY

A PRACTICAL GUIDE FOR TOXICOLOGISTS AND BIOMEDICAL RESEARCHERS

Animal Hematotoxicology

A Practical Guide for Toxicologists and Biomedical Researchers

G. O. Evans

CRC Press
Taylor & Francis Group
Boca Raton London New York

CRC Press is an imprint of the
Taylor & Francis Group, an **informa** business

CRC Press
Taylor & Francis Group
6000 Broken Sound Parkway NW, Suite 300
Boca Raton, FL 33487-2742

First issued in paperback 2019

ISBN-13: 978-1-4200-8009-4 (hbk)
ISBN-13: 978-0-367-38709-9 (pbk)

Library of Congress Cataloging-in-Publication Data

Evans, G. O.
  Animal hematotoxicology : a practical guide for toxicologists and biomedical researchers / author, G.O. Evans.
    p. ; cm.
  Includes bibliographical references and index.
  ISBN 978-1-4200-8009-4 (hardcover : alk. paper)
    1. Toxicology, Experimental. 2. Hematology, Experimental. 3. Laboratory animals. I. Title.
    [DNLM: 1. Animals, Laboratory--blood. 2. Toxicology--methods. 3. Hematologic Tests--methods. 4. Hematology--methods. 5. Predictive Value of Tests. 6. Toxicity Tests--methods. QV 602 E92a 2008]

  RA1199.E93 2008
  615.9--dc22   2008015482

Visit the Taylor & Francis Web site at
http://www.taylorandfrancis.com

and the CRC Press Web site at
http://www.crcpress.com

Contents

Preface

This book is designed to help new entrants to the fields of laboratory animal hematology and toxicology, where the challenges are stimulating and the "rules" for human hematology do not always apply. There are many good general textbooks on human hematology and toxicology, with each organization having its favorites: this book is aimed to bridge the gap between these two mainstream sciences.

Chapters are written to cover the three main blood cell types—erythrocytes, leukocytes, and thrombocytes—and potential toxic effects on these cells. A chapter on immunotoxicology has been included as this scientific discipline is closely aligned to hematology. Other chapters are devoted to preanalytical and analytical variables that affect animal studies, as these play a far more important part when interpreting data in contrast to humans, where many of these variables can be well controlled or have less physiological effect. Information has been collated from published papers, textbooks, and unpublished data: references are provided at the end of each chapter or in appendix A, where readers are provided with some key references on published reference ranges for laboratory animals.

Biological organisms cannot be understood solely by reducing them to their component parts and numbers, and toxicological science requires people with the knowledge and experience to interpret and place the hematological findings obtained from a study into an overall picture of toxicity. Hematology data from *in vivo* toxicology studies remains one of, if not the most predictive, disciplines for human risk assessment, as the same measurements made in preclinical toxicology studies can be made in early clinical trials. About 10 years ago, an experienced toxicologist described hematology as the Cinderella of toxicology. Hopefully, this book will encourage you to help Cinderella go to the ball and find her rightful place.

The Author

Following several senior positions and 13 years in health service laboratories, G. O. Evans left the post of principal clinical chemist at Birmingham Children's Hospital, England, to enter the field of laboratory animal toxicology and clinical pathology. He was head of clinical pathology at Wellcome Research Laboratories, Beckenham, England, moving to join Astra Charnwood, and later the two AstraZeneca sites in the United Kingdom as director of clinical pathology. He is now a director of A. George Owen and Company. (e-mail: agoco.melton@fsmail.net)

He was the editor of and a contributor to a number of chapters of the book *Animal Clinical Chemistry: A Primer for Toxicologists*. In addition to lecturing at various universities in the UK, he has published over 40 papers on animal clinical pathology and toxicology. Other activities have included membership on the editorial board of the journal *Laboratory Animals* for over 15 years, and he has served as scientific secretary, chairman, and now is a lifetime honorary member of the Association of Comparative Clinical Pathology, UK.

1 Introduction

The broad scope of toxicology covers the safety evaluation of drugs, foods, pesticides, and industrial chemicals, and the preclinical animal safety evaluation studies are designed to determine the effects of xenobiotics (substances foreign to an animal), biopharmaceuticals, and dietary modifications in selected laboratory animal species or by *in vitro* models. The results obtained from these studies are then extrapolated to predict potential effects and risks for other species—human and other animals—in terms of both adverse effects and determining safe limits for human and veterinary uses or environmental exposure. In addition, information relating to adverse toxic or pharmacological effects at dosages higher than the proposed therapeutic dosages may also be useful for clinicians/physicians.

In vivo animal studies are generally designed to classify toxic effects into those developing after a single dose (acute exposure) or multiple (repeated) exposures; studies with multiple exposures are of varied duration and include short-term repeat dosing (<5% of the animal's normal life span), subchronic (5 to 20% of the life span), and chronic (the greater proportion or entire expected life span, e.g., 2 years). The possible routes of compound administration include oral, intravenous, intramuscular, subcutaneous, inhalation, intraperitoneal, and by suppository. The formulations of the test compound are mainly selected based on the route of administration and the physical, chemical, and stability properties of the compound. The dose levels are chosen generally after preliminary studies where the doses are higher than the proposed use or expected exposure; studies of drugs may include several ascending dosages that are above the proposed human therapeutic dose (Timbrell, 1999; Derelanko and Hollinger, 2001; Hayes, 2007). Evidence may be sought to show that any test-related compound effects observed during the dosing period are reversible during a subsequent period when dosing has ceased (the reversibility or recovery study phase).

For *in vivo* studies, the species of laboratory animal is chosen because it is appropriate for the pharmacological action of the test compound, or there is substantial information on other xenobiotics using the chosen species. This leads to the majority of studies being carried out with rats, mice, rabbits, dogs, and nonhuman primates to a lesser extent; other species, such as the micropig and hamster, are used less commonly.

Conventionally, compounds are tested in at least two species. In some cases, the species studied may not be a relevant pharmacologic model but can provide evidence of adverse nonpharmacologic effects, e.g., antibodies tested in nonreactive species. The choice of animal strain, e.g., for the rat, is often a local decision based on a center's long-term experience and a substantial historical database.

1

Studies are sometimes described as regulatory, implying that these studies will be submitted to the appropriate regulatory authorities based on their requirements, and perhaps following initial discussions about study designs with the regulators. These studies are usually performed to conform to the requirements of good laboratory practice (GLP) (FDA, 2007; Weinberg, 1995; OECD, 1997). Preliminary or investigatory studies, where possible, should also be performed using the same operating procedures and standards to ensure the integrity of both the data and the performance of the study, although the studies may not claim to be GLP compliant.

The bioavailability and toxicity of xenobiotics involves the complex interactions of absorption, distribution, metabolism, and excretion (ADME); these interactions vary with animal species, and thus the observed toxic effects vary among species (Timbrell, 1999; Hayes, 2007), and in the majority of studies the blood cells will be subjected to exposure to the xenobiotic or its metabolites. Toxic effects may be due to the parent compound or one or more of its metabolites, and not all compounds with similar structure have the same adverse effects—this may be due to selective chemical design and differing ADME properties. *In vitro* models may not mimic the *in vivo* ADME characteristics of a compound, and toxic effects may not be observed in both; for example, a compound may not be metabolized *in vitro* to an active metabolite that causes toxicity, or in contrast, the toxicity observed *in vitro* may be due to higher compound concentrations than those achieved in *in vivo* studies.

Blood is one of the largest organs of the body, forming about 7% of total body weight (occasionally it is called the hematon). Blood can be divided into three cellular elements—erythrocytes (the erythron), leukocytes (the leukon), and the thrombocytes/platelets (thrombon). These elements will be discussed separately in the following chapters; however, there are strong interactions between these blood components and the fluid plasma in which these cells circulate. Examination of hematologic disorders is generally focused on changes of erythrocytes, leukocytes, and coagulation by the examination of blood samples, and these blood examinations can be supplemented by examinations of bone marrow and histopathology of various tissues. Hematology is an essential part of toxicology where all the scientific information can be integrated and assessed for risks to human and animal health and to the environment.

Most repeat dose studies include measurements of hematology, plasma or serum chemistry, and urinalysis as routine screening tools, rather than being selected for compounds thought to have hemotoxic potential. There are several different requirements for hematology made by regulatory authorities (EPA, 2000; OECD, 1998a, 1998b, 2000, 2002), and the basic or core measurements made in toxicology safety studies have been summarized by Weingand et al. (1992, 1996). There is little controversy about these core tests, as many laboratories routinely perform them, and the required technologies are generally available. These core hematology tests are:

Erythrocyte count (red blood cell [RBC]) count
Hemoglobin
Hematocrit (packed cell volume or erythrocyte volume fraction)

Mean corpuscular volume
Mean corpuscular hemoglobin
Mean corpuscular hemoglobin concentration
Reticulocyte count
Total and differential leukocyte (white blood cell [WBC]) count
Thrombocyte (platelet) count
Prothrombin time
Activated partial thromboplastin times

A full blood examination that includes the measurements listed above is very often sufficient in toxicology studies. Five-part differential leukocyte counts, which include counts for lymphocytes, neutrophils, monocytes, eosinophils, and basophils, are preferred to three-part differential counts, which count three leukocytic populations. These listed tests should not be regarded as exclusive, and additional tests may be used to further characterize the toxicity and its underlying mechanisms. There are some items of the regulatory requirements that continue to cause debate—these include the assessment of clotting potential, differential counts without total leukocyte counts in longer-term studies, and reticulocyte counts. For immunotoxicity, the requirements are less clear, and these are discussed in Chapter 7.

The majority of blood cells have a much shorter life span than other body cells, and a single blood sample gives a snapshot of blood where cells are rapidly leaving and being replaced at differing rates. Blood samples are generally taken at the end of the dosing period, but more frequent interim sampling can yield additional information; however, the frequency of blood sampling should be such that it does not jeopardize the health of the animal or induce major blood changes. Sampling frequency is limited by the volumes of blood that can be taken, particularly from small laboratory rodents. For nonrodent studies where more frequent blood sampling is permitted, interim samples can be taken and used as a measure of intra-animal variability and progression of toxicity. Where the study includes a reversibility phase, measurements should be made from the same animals at the ends of the dosing and reversibility periods. Blood volumes sampled should be consistent for all animals, including those in untreated or control groups. The potential effects due to blood samples taken for toxicokinetic measurements also should be considered in the study design and interpretation.

The phrase "some indication of clotting potential" used in some regulatory documents is open to interpretation (Theus and Zbinden, 1984). Platelet measurements are usually incorporated into the automated full blood count, but alone these counts do not provide an adequate assessment of impaired coagulation, which can occur without an obvious change in platelet numbers. However, the requirement of an extra citrated sample for coagulation tests (prothrombin [PT] and activated partial prothromboplastin time [APTT]) may be ruled out by the required amount of blood that can be taken from the smaller laboratory animals; in this instance, a study may have to be designed appropriately to enable coagulation tests to be made by including additional animals or omitting plasma biochemical tests that would normally accompany the sampling for core hematology tests. This blood volume limitation generally does not apply to larger animals such as the dog. The simple assessment of bleeding time, which can be conducted in the animal room, is time consuming but may yield useful information if carefully done. The test involves a needle stab to stimulate bleeding, and then observing the time taken for the blood flow to stop. Where coagulation tests such as the PT or APTT are altered, additional coagulation measurements of the coagulation cascade may be performed. Some laboratories include plasma fibrinogen routinely as part of their core coagulation tests.

It is usual to prepare blood films (slides/smears) for samples submitted for automated analysis; the examination of a blood film remains a keystone in confirming data obtained by automated counting, and can provide further information on cell morphology changes, e.g., erroneously high platelet counts due to red cell fragments, erythrocyte shape changes, leukocytosis, neutrophil or platelet aggregation, parasitic or bacterial infections, etc. These blood films provide a semipermanent record that allows microscopic examination for some considerable time after sampling, unlike whole blood samples. The films remain a useful training media for inexperienced hematologists, and most laboratories have a secure storage and archive system that allows for future examinations. It is not always necessary to examine all blood films from a toxicology study. Some laboratories examine blood films from the control and high-dose groups, and where there is a possible treatment effect, examine the other dose groups to help to determine a no-observed-effect level (NOEL). Some automated hematology analyzers generate flags in the data to indicate values that are outside preset ranges, and films are examined when these alert flags are generated. The procedures for selective examination of blood films should be covered by the standard operating procedures of the laboratory, or detailed in a study protocol/plan amendment. The knowledge and experience of the responsible hematologist is an essential part of this process.

Long-term exposure to some chemicals can cause malignant disease, and some regulatory guidelines for carcinogenicity require blood films/smears to be taken at 52 and 104 weeks for differential cell counts without making other hematological measurements. It is argued that these slides can be used to detect and indicate hematogical neoplasias by a differential count of the blood films; if this limited approach for blood smears is used, then the manual differential count can only be expressed as percentage values and not absolute values, which can be determined from a total leukocyte count. It can be advantageous to perform a full blood count and examine the film in these studies, where typically only a few animals will be affected in relatively large group sizes of 25 to 50 animals, and where there is evidence of hematological effects in previous studies.

Reticulocyte and bone marrow cytology are not included in some of the regulatory requirements, although these counts and examinations are important for characterizing erythropoietic changes. Laboratories with analyzers capable of automated counting of reticulocytes generally include this measurement in the core battery of tests. If automated reticulocyte counting is not available, blood films can be prepared and stained with supravital stains for evaluation and counting. In addition to blood measurements, hemopoiesis may by assessed by examining bone marrow aspirates or bone sections; sections of the bone marrow from the sternum or other active bone marrow sites are often taken as part of the histological examination and bone marrow films/smears prepared at necropsy for subsequent examination.

For a variety of reasons, additional hematology tests may be required in a study; these tests include methemoglobin, Heinz body counts, hemotinic tests, platelet aggregation, etc. The hemotinic markers, e.g., iron, ferritin, transferrin, erythropoietin, folate, and vitamin B_{12}, are less commonly measured in animal toxicology studies than in human medical practice; this is partly due to a lack of suitable methods and reagents, and the availability of animal histopathology data, which are not readily available in clinical practice. Hematology measurements are being used increasingly in reproductive toxicology studies, although there are no current regulatory requirements. There are demands for studies that explore and develop our knowledge of the mechanisms that cause hemotoxicity.

DEVELOPING AREAS OF INTEREST

There are future opportunities for expanding the current repertoire of tests applied in toxicology studies, for example, in studying the effects on bone marrow cells and their precursors, coagulation disorders, and hemotinics, where this can be justified as part of the drug development processes. Although core hematological measurements can be made by automated blood cell counters and coagulyzers, many of the current hematology analyzers designed for human medicine may generate upwards of 15 to 20 blood parameters, but often these measurements are either unsuitable or have not been validated for blood obtained from laboratory animals. The ingenuity of the instrument manufacturers in adding various additional measurements, e.g., red cell distribution width, platelet volume, platelet crit, reticulocytic fluorescence ratios, and flagging for cellular size and variations of cell size, is largely validated for human medicine, but requires much more careful validation before exploitation in toxicological studies with different species. The more recent general hematology analyzer capabilities for measuring human hemopoietic stem cells and immature granulocytes could lead to future developments of cell counting for animals.

In vitro tests have a role to play, as they may indicate direct toxicity of a xenobiotic by direct lysis of erythrocytes or inhibition of growth in bone marrow cell cultures. Toxicity *in vivo* may result from the action of a metabolite that is not produced by *in vitro* systems, e.g., butoxyethanol causes hemolytic anemia when given orally to rats but does not cause *in vitro* hemolysis using rat blood, whereas 2-butoxyethanol does cause erythrocytic breakdown *in vitro*, i.e., the metabolite is the causative agent. Similarly, nitroso-chloramphenicol is a metabolite of chloramphenicol that does not appear in humans, but is a stronger inhibitor than the parent chloramphenicol in mouse hemopoietic cell cultures. For all *in vitro* models, it is important to ensure that the concentrations of test compounds used are relevant to exposure and metabolism in animals.

As preclinical science develops, there is an increasing emphasis on the inclusion of efficacy or pharmacological markers in addition to those biomarkers that have been the traditionally used as markers of toxicity. Biomarkers are measurements made on body fluids, tissue, or excreta to give a quantified indication of exposure to an active xenobiotic or a change in a disease process, and these markers should be sensitive and reproducible. Although there have been some attempts to classify biomarkers, in this book the term *efficacy biomarkers* will used for those markers that are applied to monitor pharmacologic rather than toxic responses. Ideally, all biomarkers should bridge between preclinical and clinical studies.

Our understanding of the immunological effects of drugs, desired or adverse, is improving, but immunological assessment is for the most part not a current mainstream activity in toxicology studies; however, this area has a strong synergistic link to hematology.

In the future, other techniques may be used more extensively to study the progression of hematological changes: ultrasound, computed tomography, magnetic resonance imaging, endoscopy, and isotopic labeling of molecules are techniques used in specialized human clinical medicine, and some of these techniques could be particularly useful in discriminating bone marrow effects as well as other organ functions. Structural activity relationships (SARs) have probably not been thoroughly exploited in the prediction of hemotoxicity, and this offers an opportunity for the future.

Laboratory animals are not small humans, but research using animals has contributed to most areas of medical research and environmental science. *In vivo* animal studies are essential in relating distribution and metabolism of a xenobiotic to safety risk assessments (Heywood, 1990; Dayan, 1991; Greaves et al., 2004). Some of the hematological differences between species and some of the variables that affect animal studies are discussed in later chapters. Any investigator carrying out animal studies must consider the three Rs: reduction of numbers, refinement, and replacement by other techniques where possible (Rowan, 1990). Individual responsibility for performing animal studies is the "fourth R."

Animals in toxicology studies are generally purpose bred with a known genetic background, although there is some debate as to whether using heterogeneous populations would be better than the current relatively homogeneous populations in the detection of idiosyncratic metabolism and adverse effects. In today's animal houses parasitic infections are uncommon but occasionally occur and can alter the hematologic picture without parasites being observed in the blood films. Bacterial or viral infections may occur, and these are sometimes associated with tissue injury, e.g., in-dwelling intravenous sets. Very rarely intervening drug treatment for clinical reasons may be necessary during a toxicology study; some examples are treatment for infections or minor injury, and in these instances hematologic assessments may be used to monitor treatment and recovery. Aspects relating to biological and chemical safety in animal facilities must be considered when transporting and handling biological samples (Wood and Smith, 1999).

The development of transgenic animals has increased the number of animals used in drug development, as the development of models of human diseases is aimed at helping and accelerating drug design and development (Yang et al., 1995; Dunn et al., 2005). Laboratories examining samples from transgenic animals need to ensure that methods are suitable for these animals, as the hematological pictures are highly variable but will not be discussed in this book.

Parts of today's society appear to expect a guarantee that adverse effects are unlikely to occur by environmental exposure to agricultural chemicals, household chemicals, or with drugs at marketed dosages. Industry needs to educate people about the risk–benefit analysis that applies to most substances, and to ensure there are adequate risk labels and antidotes for harmful substances. It will always be difficult using conventional toxicity study designs to predict idiosyncratic reactions, i.e., an adverse effect that occurs at a very low incidence rate. For instance, if the incidence rate of life-threatening bone marrow failure with a drug occurs in 1 in 100,000 humans, then the chances of identifying this risk are small when maybe a total of 1,000 animals have been exposed in the preclinical testing phase; a much larger number of animals would be required to detect these adverse events, and this is in direct conflict with the reduction in the three Rs (Zbinden, 1973).

Toxicology and the development of new drugs or industrial chemicals are a team effort involving chemists, formulation chemists, biologists, analytical chemists, animal care personnel, toxicokinetics, pathologists, statisticians, information scientists, and more. Hematologists play a vital role in these processes, as they are key gate-keepers for human safety.

REFERENCES

Dayan, A. D. 1991. A clinical scientist's view of preclinical drug testing requirements. *Hum. Exp. Toxicol.* 10:395–97.

Derelanko, M. J., and Hollinger, M. A. 2001. *Handbook of toxicology.* 2nd ed. Boca Raton, FL: CRC Press.

Dunn, D. A., Pinkert, C. A., and Kooyman, D. L. 2005. Foundation review: Transgenic animals and their impact on the drug discovery industry. *Drug Discov. Today* 10:757–67.

EPA. 2000. Toxic substances control act test guidelines: Final rule. 40 CFR. Part 799. *Fed. Reg.* 65:78745–879.

FDA. 2007. *Good laboratory practice for non-clinical laboratory studies.* Title 21 CFR 58. Electronic Code of Federal Regulations. Food and Drug Administration.

Greaves, P., Williams, A., and Eve, M. 2004. First dose of potential new medicines to humans: How animals help. *Nat. Rev. Drug Discov.* 3:226–36.

Hayes, A. W., ed. 2007. *Principles and methods of toxicology.* 5th ed. Boca Raton, FL: CRC Press.

Heywood, R. 1990. Clinical toxicity—Could it have been predicted? In *Animal toxicity studies: Their relevance for man*, ed. C. E. Lumley and S. R. Walker, 57–67. CRM Workshop series. London: Quay Publishing.

OECD. 1997. *Principles of good laboratory practice and compliance montoring.* Paris: Organization for Economic Co-operation and Development.

OECD. 1998a. *Guidelines for the testing of chemicals. Test 408. Repeated dose 90-day oral toxicity study in rodents.* Paris: Organization for Economic Co-operation and Development.

OECD. 1998b. *Guidelines for the testing of chemicals. Test 409. Repeated dose 90-day oral toxicity study in non-rodents.* Paris: Organization for Economic Co-operation and Development.

OECD. 2000. *Guidance notes for analysis and evaluation of repeat-dose toxicity studies.* Series on Testing and Assessment 32 and Series on Pesticides 10. Paris: Organization for Economic Co-operation and Development.

OECD. 2002. *Guidance notes for analysis and evaluation of chronic toxicity and carcinogenicity.* Series on Testing and Assessment 35 and Series on Pesticides 14. Paris: Organization for Economic Co-operation and Development.

Rowan, A. N. 1990. Refinement of animal research technique and validity of research data. *Fundam. Appl. Toxicol.* 15:25–31.

Theus, R., and Zbinden, G. 1984. Toxicological assessment of the hemostatic system, regulatory requirements and industry practice. *Reg. Toxicol. Pharmacol.* 4:74–95.

Timbrell, J. A. 1999. *Principles of biochemical toxicology.* 3rd ed. London: Taylor & Francis.

Weinberg, S., ed. 1995. *Good laboratory practice regulations*. New York: Marcel Dekker.

Weingand, K., et al. 1992. Clinical pathology testing recommendations for nonclinical and safety studies. *Toxicol. Pathol.* 20:539–43.

Weingand, K., et al. 1996. Harmonization of animal clinical pathology testing in toxicity and safety studies. *Fundam. Appl. Toxicol.* 29:198–201.

Wood, M., and Smith, M. W. 1999. *Health and safety in laboratory animal facilities*. Laboratory Animal Handbook 13. London: Royal Society of Medicine Press.

Yang, B., Kirby, S., Lewis, J., Detloff, P. J., Maeda, N., and Smithies, O. 1995. A mouse model for β-thalassemia. *Proc. Natl. Acad. Sci. USA* 92:11608–12.

Zbinden, G. 1973. *Progress in toxicology: Special topics*. Vol. 1. New York: Springer-Verlag.

2 Hemopoiesis, Blood Cell Types, and Bone Marrow Examination

HEMOPOIESIS

Blood contains a number of cellular elements, including the erythrocytes (red blood cells), which contain the hemoglobin important for oxygen transport; the leukocytes (white blood cells); and thrombocytes (platelets); these cells circulate within the fluid plasma. The term *hemopoiesis* or *hematopoiesis* is used to describe the formation of blood: *hemo* or *hemato* meaning "blood" and "*poiesis*" to make.

The production, differentiation, and maturation processes of blood cells are complex, involving many cell types and both local and systemic growth factors. There are a limited number of hemopoietic stem cells that are pluripotent, and these develop via the progenitor cells into the mature blood cells. Hemopoiesis is observed in the yolk sac during the fetal period with recognizable erythrocytic precursors present in early embryonic stages, and with evidence of leukopoiesis and thrombopoiesis occurring at a slightly later stage (Lajtha, 1970; Haig, 1992; Testa and Molineux, 1993; Foucar, 2001; Bain, 2006). During gestation the liver and spleen become increasingly important, and later normal hemopoiesis is controlled by the bone marrow with the activity primarily in the central skeleton and proximal ends of the long bones of humans, dogs, and rabbits, with fat mainly filling the remainder of the marrow. In rats and mice, most of bone medullary space is occupied by hemopoietic tissue, and splenic hemopoiesis plays a much more prominent role in these laboratory animals compared to dogs and humans, where increased hemopoietic activity is usually provided by the medullary space except in extreme circumstances. The hemopoietic microenvironment or tissues are an integrated network of cells that initiate and provide continual cell differentiation cycles from the small germinal cell population (the stem cells). The liver and spleen may resume their extramedullary hemopoiesis, where there is a marrow replacement, e.g., in myelofibrosis or with increased demand in severe hemolytic anemias.

In addition, the spleen and lymph nodes are involved in the interaction of lymphocyte subpopulations (together with macrophages), which enable the presentation and recognition of antigens, cell-mediated killing, and antibody production. Within the spleen, a proportion of the circulating blood volume passes from the arterial to venous circulation via the splenic sinuses and intraendothelial slits, which effectively act as a sieve trapping the opsonized bacteria, larger erythroid inclusions, e.g., Heinz bodies and Howell–Jolly bodies; the spleen also sequesters some of the senescent erythrocytes, leukocytes, and platelets. Contraction effects of the splenic muscular wall are variable between species, and contractions may be stress induced; this splenic contraction can lead to erythrocytosis in the dog. So effectively, the spleen has several functions: phagocytosis, lymphopoiesis, hemopoiesis, and storage and release of blood cells, particularly erythrocytes.

It was in studies with irradiated mice that certain bone marrow cells were observed to have the ability to form hemopoietic colonies: later it was recognized that marrow stem cells give rise to all types of blood cells, and these cells are therefore termed pluripotent or colony-forming units (CFUs). Common primitive stem cells of the bone marrow have the capacity to self-replicate, proliferate, and differentiate to increasing specialized progenitor cells, which after cell divisions within the bone marrow, result in the formation of mature cells: erythrocytes, granulocytes, monocytes, platelets, eosinophils, and basophils (Figure 2.1).

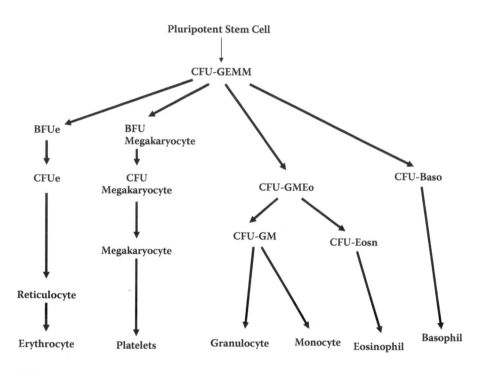

FIGURE 2.1 Simplified hemopoietic cell lineages.

The hemopoietic microenvironment can be divided into three sequential compartments: the stem cell compartment, the progenitor cell compartment containing the CFUs, and the precursor cell compartment. The stem cell compartment contains hemopoietic stem cells (HSCs). The cells of the precursor compartment may be further subdivided into proliferating or mitotic cells, maturation and storage pools, but there are no exact discrete anatomic locations for these compartments and subcompartments. The abbreviations and pathways used by different authors to describe the cell lineages, as well as CFUs and cell factors, are varied; some of the common abbreviations used are:

Baso or B	Basophilic
CFU	Colony-forming unit
E	Erythroid
Eosn or E	Eosinophilic
EPO	Erythropoietin
GEMM	Granulocytic/erythroid/monocyte/megakaryocytic
GM	Granulocytic/monocytic
L or LM	Lymphoid
M	Monocytic
SCF	Stem cell factor
TPO	Thrombopoietin

The stem cells can differentiate into CFUs that are capable of developing into myeloid cell types or the colony-forming unit granulocyte/erythroid/monocyte/megakaryocyte (CFU-GEMM), which then develops into the more defined progenitor cells, e.g., colony-forming unit granulocyte/monocyte (CFU-GM) and colony-forming unit erythroid (CFU-E). Some stem and progenitor cells may also circulate in the peripheral blood.

Hemopoiesis is closely regulated by glycoprotein growth factors produced by stromal cells, T-lymphocytes, the liver, and the kidney for erythropoietin; some growth factors act mainly on surface receptors of the primitive cells, while others act later on cells committed to a particular lineage. These glycoproteins are divided into two main classes: the colony-stimulating factors and the interleukins (ILs) (Schwarzmeier, 1996). The binding of a growth factor with its surface receptors leads to the activation of complex biochemical pathways by which messages are transmitted to the nuclei. These pathways involve the sequential phosphorylation of substrates by protein kinases that in turn activate transcription factors, which then either activate or inhibit gene transcription. Some of these growth factors inhibit apoptosis (programmed cell death) of their target cells. Stem cells, CFU-GEMM, and CFU-L are capable of self-renewal under normal circumstances, but the potential for self-renewal of cells declines with differentiation.

BONE MARROW STROMA

The bone marrow has a network of venous sinuses that drain into the central venous vessel, nutritive vessels, and capillaries, with hemopoiesis occuring in the extravascular spaces between these marrow sinuses. The stromal cells are a mixture of endothelial, fibroblast, adipocyte, and macrophage cells that secrete cytokines and synthesize components of the extravascular matrix, and these cells have adhesion molecules that react with corresponding ligands on the stem cells. The luminal layer of sinus wall endothelial cells allows mature blood cells to pass through transient migration pores, and various substances into the extravascular space. The term *diapedesis* is occasionally used to describe the movement of reticulocytes into the marrow sinusoids.

ERYTHROCYTES

The erythron is composed of erythrocytes and their precursors located in the blood vessels, bone marrow, or extravascular sites. The earliest recognizable precursor is the pronormoblast/rubriblast, which progressively matures to the reticulocyte and circulating erythrocyte (Figure 2.2). The erythroid series are derived from the stem cells via the CFU-GEMM, then the burst-forming unit erythroid (BFU-E), to give CFU-E cells, which are restricted to erythrocytic cell lineage. The last stage of maturation is the reticulocyte—a nonnucleated biconcave erythrocyte with a fine basophilic reticular network, which retains ribose nucleic acid (RNA). The developing red cell in the bone marrow (erythroblast) is nucleated, but the nucleus condenses with maturation, prior to extrusion from the bone marrow and release into the circulation. About 10 to 15% of developing erythroblasts die within the bone marrow and therefore do not develop into mature red cells. Thus, within six cell divisions, the pronormoblast develops into the anucleate immature reticulocyte, which has a life span of about 2 days after release into the circulation.

Some of the principal stimulatory growth, inhibitory, and transcription factors for erythropoiesis are given in Table 2.1.

Erythropoietin produced in the peritubular complex of the kidney, liver, and other organs stimulates mixed cell lineage and erythrocytic progenitor cells, pronormoblasts, and early erythroblasts to proliferate.

When oxygen delivery to the tissues is reduced as a consequence of reductions in circulating red cell mass, renal synthesis of erythropoietin (EPO) increases. Increased levels of EPO then bind to receptors for BFU-E and CFU-E in the bone

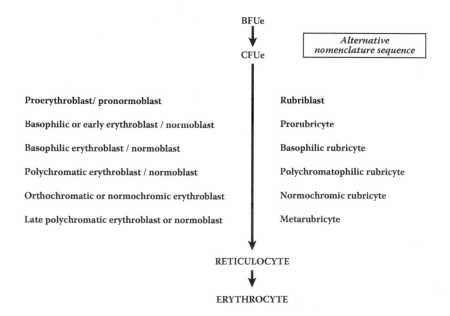

FIGURE 2.2 Erythropoietic cell lineage.

TABLE 2.1

Stimulating Growth Factors	Inhibiting Factors
Erythropoietin	TNF-α
Interleukins 3, 6, and 11	Interleukin 1β and IL-2
GM-CSF	Interferon γ
Insulin-like growth factor 1	Transforming growth factor β
Hepatocyte growth factor	MIP 1α
Thrombopoietin	Glucocorticoids
Testosterone	
Kit ligand, stem cell factor	

marrow, and cause a shortening of their cell cycle times, which results in increased maturation and the release of red cells from the bone marrow. As a consequence of the now increased circulating red cell mass, tissue oxygenation improves and there is feedback mechanism onto EPO production.

An intracellular mean cell hemoglobin concentration (MCHC) of approximately 35 g/dl appears to be an important factor that limits cell divisions, and this MCHC value of approximately 35 g/dl is found across all of the healthy adult common laboratory animals, whereas other erythrocytic parameters differ more between species. Erythrocytes contain hemoglobin, with a switch from fetal to adult hemoglobin forms occurring during the neonatal period.

LEUKOCYTES

Production of granulocytes and monocytes occurs in the bone marrow, and the promyelocytes are the earliest recognizable precursors. Thus, this development goes from CFU-GEMM to CFU-GM via the promyelocytes, which undergo further cell division and mature into the myelocytes, metamyelocytes, and finally the granulocytes—neutrophils, eosinophils, and basophils (Figure 2.3). The production of granulocytes and monocytes is regulated by the growth factors interleukin 3 (IL-3) and the granulocyte/macrophage colony-stimulating factor (GM-CSF), and also acts on CFU-GEMM to produce CFU-GMEosn and CFU-Baso.

Granulocytopoiesis has two main phases: proliferation and maturation. The mitotic pool of cells—the myeloblasts, promyelocytes, and myelocytes—undergoes three to five divisions before yielding the postmitotic pool of metamyelocytes, band cells, and mature segmented cells. The postmitotic cells are distributed into four compartments: the maturation, storage, circulating, and tissue pools. The maturation pool contains metamyelocytes and band cells that mature but do not proliferate. In the storage pool, the cells include segmented neutrophils that act as a reservoir of mature cells, particularly in dogs and cats; these cells can be released into the circulation at short notice, e.g., as a response to stress. Within the circulating pool, the granulocytes can be located either in a freely circulating pool or within the marginal pool from where cells leave the circulation. Mature cells enter the circulation usually for a short period before migrating to the tissues and body cavities, and many cells

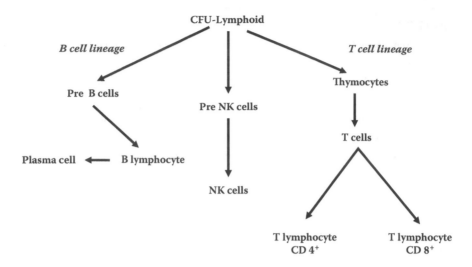

FIGURE 2.3 Lymphocytic lineage.

stay for only a short time within the tissue pool, limited by their active function and before consumption by the tissue. The spleen, liver, and lymph nodes in most species are able to produce granulocytes when there is increased demand—this is termed extramedullary granulocytopoiesis. Leukocytes are less numerous than erythrocytes and platelets in blood.

Collectively the neutrophils, eosinophils, and basophils are termed granulocytes because of their granular contents. Granulocytes are also know as the polymorphonuclear leukocytes or polymorphs, having segmented nuclei and cytoplasmic granules. These cells have a short life span and a rapid turnover.

LYMPHOCYTES

Lymphopoiesis differs from other hemopoietic pathways that have a biphasic process of differentiation followed by maturation; the lymphoid stem cell differentiates to form mature antigen committed lymphocytes in the primary lymphoid organs with T lymphocyte differentiation occurring in the thymus gland and B lymphocyte differentiation occurring in the fetal liver and then the adult bone marrow (Figure 2.3). Antigen-dependent proliferation and development of mature T lymphocytes (T cells) and B lymphocytes (B cells) take place in the secondary lymphoid organs: spleen, lymph nodes, and mucosa-associated lymphoid tissues.

The mature lymphocytes are small mononuclear cells derived from hemopoietic stem cells; a common lymphoid stem cell undergoes differentiation and proliferation to give rise to two major subpopulations—the B cells and T cells. B cell lymphocyte maturation occurs primarily in the bone marrow, and these cells mediate antibody immunity or humoral responses. T cells mature primarily in the thymus and also involve the lymph nodes, liver, spleen, and parts of the reticuloendothelial system (RES), and T cells play a vital role in cell-mediated immunity. Both the receptor on

T cells (TCR) and surface membrane immunoglobulin (sIG) on B cells are molecules that have constant and variable regions. Mature T cells can be subdivided into three main types: the helper cells expressing CD4 antigen, the suppressor cells expressing CD8 antigen, and the natural killer/cytotoxic cells, which can also express CD8.

The antigens expressed on the surface of a cell can now be recognized in the laboratory by using monoclonal antibodies, and this has led to the classification of these antigens, which is known as the cluster of differentiation (CD) nomenclature system. Although the CD nomenclature is most frequently applied to lymphocytes (see later chapters and appendices), the nomenclature system now incorporates other hemopoietic cell markers.

Immune response specificity is dependent on the amplification of antigen-selected T and B cells, and results from the interaction of the T cells, B cells, and antigen presenting cells (APCs). The immune system also involves proteins from immunoglobulins (IgG, IgM, IgA, IgE, and IgD) and complement protein families (C3b, C3, and C5a) (see Chapter 7). Natural killer (NK) cells are often large granular lymphocytes with prominent granules, and they are neither T nor B cells, although they may be CD8 positive. These NK cells can kill target cells by direct adhesion or bind to a target cell that has bound antibody (termed antibody-dependent mediated cytotoxicity [ADCC]).

PLATELETS

Megakaryocytes are large multinucleated cells derived from the megakaryoblast developed from hemopoietic stem cells in the bone marrow (Figure 2.4). The platelets are formed by breaking off from the megakaryocyte cytoplasm and then enter the bloodstream. The platelets are nonnucleated and appear in peripheral blood films in granular basophilic forms. The hormone thrombopoietin stimulates megakaryocyte numbers by increasing differentiation of stem cells into megakaryocytes, increasing the number of megakaryocyte divisions (ploidy), and thus platelet production.

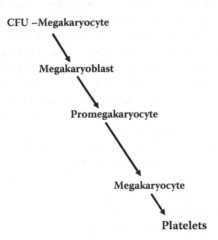

CFU –Megakaryocyte

Megakaryoblast

Promegakaryocyte

Megakaryocyte

Platelets

FIGURE 2.4 Thrombopoietic cell lineage.

Both thrombopoietin and erythropoietin act on the mixed-lineage progenitor cells, although their respective dominant actions are on platelets and red cells.

Platelets are the second most numerous cells in blood and circulate for about 6 to 8 days in dogs (4 to 5 days in rodents), and then are destroyed by the spleen or in the pulmonary vascular bed. They have major roles in hemostasis, triggering coagulation and by their aggregation to prevent bleeding (see Chapter 8). Much emphasis is placed in the literature on the myeloid and erythroid series, but thrombocytopenias are more commonly associated with drug effects, so megakaryocytes may receive greater attention in the future.

PLASMA

Blood cells circulate in plasma, which occupies about 50 to 60% of the total blood volume, and the plasma contains electrolytes, proteins, and other small molecules utilized and produced by the various organs and tissues. The plasma proteins of particular interest to hematologists are:

Hemostatic proteins involved in coagulation and fibrinolysis
Immunoglobulins
Complement proteins
Transport proteins involved with erythrocytes and iron metabolism

MONOCYTE-MACROPHAGE SYSTEM (OR RETICULOENDOTHELIAL SYSTEM)

This system is formed by monocyte-derived cells and includes the Kupffer cells of the liver, alveolar macrophages in the lung, mesangial cells in the kidney, microglial cells in the brain, and macrophages within the bone marrow, spleen, lymph nodes, skin, and serosal surfaces. These cells are particularly localized in tissues that may come into contact with external pathogens and allergens, and the system enables cells to communicate with the lymphoid cells and cells within the liver, spleen, lymph nodes, bone marrow, thymus, and intestinal tract–associated lymphoid tissues. The functions of the monocyte-macrophage (or RES) system include phagocytosis and destruction of pathogens and cell debris, processing and presentation of antigens to lymphoid cells, and the production of cytokines involved in the regulation of growth factor and cytokine networks governing hemopoiesis, inflammation, and cellular responses. The antigen presenting cells (APCs) react principally with T cells in the spleen, lymph nodes, thymus bone marrow, and other tissues. An antigen-specific monocyte migration factor is produced by lymphocytes, and this causes monocytes to move and remain in areas where antigens are concentrated.

The spleen plays an important role in the filtering of blood from the arteriolar circulation—the white pulp—through the endothelial mesh of the red pulp to the venous sinuses. During these filtration processes, effete cells, unwanted material from deformable red cells (hemosiderin granules, nuclear remnants), and particulate matters (e.g., opsonised bacteria) are removed.

Following chapters discuss separately the blood cell types, their functions, and potential toxic effects on these cells. In the appendices, several references are given for publications that contain microphotographs of blood to which the reader can refer, but these cannot replace the examination of blood films as part of the education process. These publications use different stains and microscope magnifications to demonstrate particular cell characteristics to the best advantage; in the local laboratory, the staining procedures and microscopic lens magnifications may differ from those in these publications.

In hemotoxicity, chronic changes of cell populations rarely occur alone, and this can be explained by a cursory understanding of the development of blood cells from the earlier cell lineages; thus, the term *pancytopenia* is used to describe these combined effects on erythrocytes, leukocytes, and thrombocytes.

BONE MARROW EXAMINATIONS

Various techniques have been described for the preparation and examination of bone marrow smears, and success using these techniques depends on the procedures for collection, smear preparation, staining, and careful microscopic evaluation or flow cytometry. Given manual and flow cytometry techniques are labor intensive, many laboratories elect to examine bone marrow smears when there is evidence suggesting hematopoietic effects from either the hemograms or histological evidence from bone marrow sections and examination of the spleen and other tissues. It is common practice for bone marrows to be sampled from subchronic and chronic studies, but only examined at the discretion of the study director, pathologist, or hematologist.

A biopsy specimen contains trabecular bone—a meshwork of bony plates and strands containing the hemopoietic tissue and stroma—and the dense cortical bone. There is a variety of cells in both hemopoietic and nonhemopoietic tissue. Within the stroma there is a mixture of cells and binding proteins, including fibroblasts, adipocytes, fibronectin, proteoglycans, and macrophages. Bone marrow cellularity varies with stress (Moeller, 1991), site, and age, and it is subject to diurnal variation (Clark and Korst, 1969).

BONE MARROW SMEAR PREPARATION

Bone marrow samples should be collected very shortly after or within a few hours of death (Andrews, 1991; Smith et al., 1993; Valli et al., 1990; Matsumoto, 1991; Bollinger, 2004). For rats and mice, the femoral bone marrow is commonly used for examination; in dogs, bone marrow sample is usually taken from the iliac crest, sternum, or a rib, as the femoral marrow cavity has a high fatty content. The midshaft of the long bones is generally fatty in larger animals, and the marrow should be collected from the cellular marrow at the ends of these bones, or the sample should be taken from the sternum or a rib.

After cutting the ends of rat and mouse bones, the contents of the bone may be flushed through with measured volumes of saline or other suitable fluids, e.g., Hank's balance salt solution with albumin. Alternatively, bone marrow can be transferred from the cut ends by a camel-hair brush or spatula onto the glass slide. Several laboratories mix the marrow with a solution of bovine albumin in physiological strength saline (Berenbaum, 1956) or a pooled plasma for the species under investigation prior to preparing the smear. Some investigators mix an anticoagulant (EDTA) with the bone marrow sample (Valli et al., 1990). Dilutions of bone marrow with solutions should be carefully controlled to maintain consistent intersample comparisons. Cross contamination between samples should be avoided when using a glass slide or a camel-hair brush to draw the marrow across a glass slide for examination. The staining and thickness of smears is rarely uniform, with thicker and unreadable smears occurring more often than smears that are too thin.

The prepared slides should be allowed to dry and then stained. The marrow smears may be stained with Giemsa- or Romanowsky-based stains. Perl's Prussian blue stain may be applied where there are perturbations of iron metabolism, and other cytochemicals can be used to differentiate abnormal or immature cells.

The cytologic and histologic evaluation of bone marrow should include comments on the degree of cellularity and cell typing. Several suitable areas of the smear should be examined for satisfactory microscopic examination. Given the lack of uniformity, examinations should not rely on a single field, and some cell smears may be too thick for examination. A total of at least 200 cells should be counted.

The initial reporting step is usually to estimate the proportions of myeloid (granulocytic or M) to erythroid (E) lineage cells and express these as the M:E ratio. Alterations in the M:E ratio can be caused by a myeloid hyperplasia or an erythroid hypoplasia, or vice versa; thus, the ratio may appear normal in situations where both the M and E constituents are simultaneously up- or down regulated. The reporting of the M:E ratios should always be accompanied by some additional observations of the marrow cellularity, some morphological assessment of precursor cells, and sample quality. It is less common to classify all identifiable cells by lineage and developmental stages, i.e., to perform a full myelogram. Additional indices for the relationship between myeloid and erythroid cell lineages may be also expressed, e.g., maturation indices for erythroid (EMI) and myeloid (MMI) cells and myeloid and erythroid left-shift indices (Brown, 1991). Published data for M:E ratios for laboratory animals are variable and dependent on the analytical method.

Flow Cytometric Examination of Bone Marrow Aspirates

Flow cytochemicals with immunological markers can be used for the identification and counting of bone marrow cells (Terstappen and Levin, 1992; Martin et al., 1992; Criswell et al., 1998, Saad et al., 2000; Schlucter et al., 2001; Brott et al., 2003; Weiss, 2004). Bone marrow samples for flow cytometry are collected in a manner similar to that of those samples used for manual counts; the mixing of bone marrow and detection reagents should be performed shortly after the bone marrow collection to avoid the effects of rapid cell degeneration.

Refer to your local laboratory for their preferred methods for bone marrow examinations or consult with other laboratories about the procedure.

In Vitro Techniques

Techniques involving bone marrow cultures are used increasingly to screen and compare compounds prior to preclinical *in vivo* studies (Deldar and Stevens, 1993; Parchment et al., 1993; Parchment, 1998). The marrow cells are cultured in media containing varied mixtures of cytokines, erythropoietin, stem cell factors, bovine albumins, etc., to promote and maintain cell growth; at the end of the incubation period, cultured cells are identified and counted using either manual or flow cytometric techniques (Brott et al., 2003). The time period required for the development of erythroid colony-forming units (CFU-E), blast-forming units (B-CFU), and granulocyte/monocyte colony-forming units (CFU-GM) varies between species. Short-term bone marrow cultures are used generally to assess potential toxicities, but longer-term cultures that mimic the hemopoietic environment in which pluripotent stem cells can develop are perhaps more useful for understanding effects on stromal cell layers (Dexter et al., 1977; Dexter and Spooner, 1987; Chang et al., 1989; Caro, 1991; Coutinho et al., 1993). *In vitro* screening methods using human umbilical cord blood, placental blood, or umbilical vein endothelial cells (HUVECs) have also been used to screen for hemotoxicants.

There is an advantage in using different cell assays for different toxic effects. Progenitor and precursors cells are suitable for acute toxic effects, stem cells for chronic toxicity, and stromal cells or microenvironmental assays for stromal effects. Some limitations of *in vitro* assays are: (1) species differences do occur, so the choice of species used for bone marrow cultures may be critical (Irons et al., 1995), and (b) toxicity may not be due to the test compound but to a metabolite that is not produced in the culture. An advantage of *in vitro* techniques is the more rapid demonstration of the reversibility of toxic effects (Yoshida and Yoshida, 1990).

TOXIC EFFECTS

Broad indications of bone marrow toxicity are obtained from peripheral blood film examinations, and these may be manifested as anemias, neutropenias, thrombocytopenias, or pancytopenias. In hypoplastic marrows where single progenitor cell lineages are affected, these cells will be absent or the maturation of cells will appear to be arrested. In very severe hypoplasia, myeloid (granulocytic) or erythroid cell lineages will be reduced or absent, but other cells—lymphocytes, plasma cells, and reticuloendothelial macrophages—will be present. With cytostatic compounds, the severity and recovery times show considerable variation. In contrast, malignant disorders may be characterized by uncontrolled clonal proliferations of hemopoietic cells, and these can be described as acute leukemias, chronic leukemias, nonleukemic lymphoproliferative disorders, and nonleukemic myelolymphoproliferative disorders.

The stromal layer promotes cell development in localized areas of the marrow, and there are strong interactions between the stromal cells and hematopoietic cells that have particularly marked effects on cell differentiation (Clark and Keating, 1995; Guest and Uetrecht, 2000). Toxic changes due to xenobiotics affecting the structure and function of cartilage or bone can occur at any age, and changes during the pre- and postnatal periods are of special interest in reproductive toxicology. Bone mineral metabolism is affected by both nutrition and hormonal mechanisms, and bone marrow studies are essential, with drugs aimed at bone remodeling.

Some points for consideration when interpreting bone marrow examinations are:

1. Quality of marrow smears
2. Cell counting variability
3. Is there evidence of marrow toxicity in the peripheral blood?
4. Is there evidence of altered bone structure in the histological bone section?
5. Are the effects limited to one or more than one cell lineage?
6. Are the effects secondary to other organ toxicities or stress?
7. Does the marrow show a recovery response?
8. Is there stromal cell toxicity?

Bone marrow examinations lead to improved understanding of some marrow toxicities, and bone marrow cell measurements can be used to screen and distinguish between similar compounds, or to show reversibility of effects.

REFERENCES

GENERAL

Bain, B. 2006. *Blood cells. A practical guide.* 4th ed. Oxford: Blackwell Publishing.
Clark, B. R., and Keating, A. 1995. Biology of bone marrow stroma. *Ann. N.Y. Acad. Sci.* 770:70–78.
Eaves, C. J., and Eaves, A. C. 1997. Stem cell kinetics. *Balliere's Clin. Haematol.* 10:233–57.
Foucar, K. 2001. *Bone marrow pathology.* 2nd ed. Chicago: ASCP Press.
Guest, I., and Uetrecht, J. 2000. Drugs toxic to the marrow that target the stromal cells. *Immunopharmacology* 46:103–12.
Haig, D. M. 1992. Haemopoietic stem cells and the development of the blood cell repertoire. *J. Comp. Pathol.* 106:121–36.
Irons, E. D., ed. 1985. *Toxicology of the blood and bone marrow.* New York: Raven Press.
Lajtha, L. G. 1970. Kinetics of haemopoiesis. *Br. J. Radiol.* 44:519.
Schwarzmeier, J. D. 1996. The role of cytokines in haematopoiesis. *Eur. J. Haematol.* 57(Suppl.):69–74.
Testa, N. G., and Molineux, G., eds. 1993. *Haematopoiesis: A practical approach.* Oxford: IRL Press.
Weiss, L. P. 1995. The structure of hematopoietic tissues. In *Blood: Principles and practice of hematology,* ed. R. I. Handin, S. E. Lus, and T. P. Stossel. Philadelphia: JB Lippincott.

BONE MARROW EXAMINATION

Andrews, C. M. 1991. The preparation of bone marrow smears from femurs obtained at autopsy. *Comp. Haematol. Int.* 1:229–32.
Berenbaum, M. C. 1956. The use of bovine albumin in the preparation of marrow and blood films. *J. Clin. Pathol.* 9:381–83.
Bollinger, A. P. 2004. Cytological evaluation of bone marrow in rats: Indications, methods, and normal morphology. *Vet. Clin. Pathol.* 33:58–67.
Brown, G. 1991. The left shift index: A useful guide to interpretation of marrow data. *Comp. Haematol. Int.* 1:106–11.

Clark, R. H., and Korst, D. R. 1969. Circadian periodicity of bone marrow, mitotic activity and reticulocyte counts in rats and mice. *Science* 166:236–37.

Criswell, K. A., Bleavins, M. R., Zielinski, D., and Zandee, J. C. 1998. Comparison of flow cytometric and manual bone marrow differentials in Wistar rats. *Cytometry* 32:9–17.

Martin, R. A., Brott, D. A., Zandec, J. C., and McKeel, M. J. 1992. Differential analysis of animal bone marrow by flow cytometry. *Cytometry* 13:638–43.

Matsumoto, K. 1991. Studies of bone marrow cells in experimental animals: Bone marrow testing in the safety study. *Exp. Anim.* 40:17–26.

Moeller, T. A. 1991. Investigations of myelotoxic effects in rats. *Arch. Toxicol.* 14(Suppl.):83–88.

Saad, A., Palm, M., Widell, S., and Reiland, S. 2000. Differential analysis of rat bone marrow by flow cytometry. *Comp. Haematol. Int.* 10:97–107.

Schlueter, A. J., Bhatia, S. K., Xiang, L., Tygrett, L. T., Yamashita, Y., de Vries, P., and Waldschmidt, T. J. 2001. Delineation among eight major hemopoietic subsets in murine bone marrow using a two-color cytometric technique. *Cytometry* 43:297–307.

Smith, C. A., Andrews, C. M., Collard, J. K., Hall, D. E., and Walker, A. K., eds. 1993. *Color atlas of comparative diagnostic and experimental hematology.* London: Wolfe Publishing, Mosby-Year Book Europe Ltd.

Terstappen, L. W. M. M., and Levin, J. 1992. Bone marrow cell differential counts obtained by multidimensional flow cytometry. *Blood Cells* 18:311–30.

Valli, V. E., Villeneuve, D. C., Reed, B., Barsoum, N., and Smith, G. 1990. Evaluation of blood and bone marrow, rat. In *Monographs on pathology of laboratory animals*, ed. T. C. Jones, J. M. Ward, U. Mohr, and R. D. Hunt, 27–33. Berlin: Springer-Verlag.

Weiss, D. J. 2004. Evalution of canine bone marrow proliferative disorders by use of flow cytometry analysis of CD45 expression and intracytoplasmic complexity. *Comp. Clin. Pathol.* 13:51–58.

In Vitro Assays

Brott, D. A., Maher, R. J., Parrish, C. R., Richardson, R. J., and Smith, A. K. 2003. Flow cytometric characterization of perfused bone marrow cultures: Identification of the major cell lineages and correlation with the CFU-GM assay. *Cytometry* 53A:22–27.

Caro, J. 1991. Clonal assays for hemopoietic progenitor cells. In *Hematology*, ed. W. J. Williams, E. Beutler, A. J. Ersley, and M. A. Lichtman. 4th ed. New York: McGraw-Hill.

Chang, G., Morgenstern, G. R., Coutinho, L. H., Scarffe, J. H., Carr, T., Deakin, D. P., Testa, N. G., and Dexter, T. M. 1989. The use of bone marrow culture for autologous bone marrow transplantation in acute myeloid leukaemia: An update. *Bone Marrow Transplant.* 4:5–9.

Coutinho, L. H., Gilleece, M. H., de Wynter, E., and Will, A. 1993. Clonal and long-term culture using human bone-marrow. In *Haematopoiesis: A practical approach*, ed. N. G. Testa and G. Molineux. 75–106. Oxford: IRL Press.

Deldar, A., and Stevens, C. E. 1993. Development and application of in vitro models of hematopoiesis to drug development. *Toxicol. Pathol.* 21:231–40.

Dexter, T. M., Allen, T. D., and Lajtha, L. G. 1977. Conditions controlling the proliferation of haematopoietic stem cells in vitro. *J. Cell. Physiol.* 91:335–44.

Dexter, T. M., and Spooner, E. 1987. Growth and differentiation in the haemopoietic system. *Ann. Rev. Cell. Biol.* 3:423–41.

Irons, R. D., Le, A. T., Som, D. B., and Stillman, W. S. 1995. 2'3'-dideoxycytidine-induced
    thymic lymphoma correlates with species-specific suppression of a subpopulation of
    primitive hematopoietic progenitor cells in mouse but not rat or human bone marrow.
    J. Clin. Invest. 95:2777–82.
Parchment, R. E. 1998. Alternative testing systems for evaluating non-carcinogenic hemato-
    logic toxicity. *Environ. Health Perspect.* 106 (Suppl. 2):541–47.
Parchment, R. E., Huang, M., and Erickson-Miller, C. 1. 1993. Roles for in vitro myelotoxic-
    ity tests in preclinical drug development and clinical trial planning. *Toxicol. Pathol.*
    21:241–50.
Yoshida, Y., and Yoshida, C. 1990. Reversal of azidothymidine-induced bone marrow sup-
    pression by 2',3'-dideoxythymidine as studied by hemopoietic clonal culture. *Aids Res.
    Hum. Retroviruses* 6:929–32.

3 Erythrocytes, Anemias, and Polycythemias

A simplified description of erythropoiesis is given in Chapter 2. In this chapter, some of the erythrocytic measurements and morphological comments, functions, and kinetics are discussed, followed by notes on various anemias and polycythemias. Preanalytical and analytical variables affecting erythrocytic measurements are discussed later in Chapters 9 and 10.

The common abbreviations for red cell measurements are:

Hb or Hgb	Hemoglobin concentration
HCT	Hematocrit (expressed as a decimal fraction or ratio); derived as HCT = RBC × MCV divided by 100; may also be measured by centrifugation as packed cell volume (PCV expressed as %)
MCH	Mean corpuscular (or cell) hemoglobin; absolute hemoglobin concentration per erythrocyte; derived as MCH (pg) = Hb divided by RBC multiplied by 10
MCHC	Mean cell hemoglobin concentration; relative quantity of Hb per erythrocyte; measured directly from the optical properties of the cell, or derived as MCHC (g/dl) = Hb divided by hematocrit
MCV	Mean corpuscular (or cell) volume; measured directly or derived as HCT multiplied by 1000 divided by RBC
RBC	Red blood cell count (erythrocyte)

The calculations are based on SI units and the expression of hemoglobin concentration as g/dl (see Appendix D). For MCV, MCHC, and MCH, the values are means and the distribution of these values is often narrow, but these distributions should be examined for any changes not shown in these mean values. These measurements indicate the number and size of erythrocytes and the levels of hemoglobin.

An automated full blood count will provide all six of these measurements, i.e., Hb, RBC, HCT, MCH, MCHC, and MCV. Other measurements include:

CHMC   Cell hemoglobin mean concentration value

HDW   Hemoglobin distribution width, which is the standard deviation
            of the hemoglobin concentration distribution

NRBC   Nucleated red blood cells

RDW   Red cell distribution width; this is a measure of variability of cell
            size, e.g., indicating anisocytosis, where the normal-sized cell
            population is altered by the presence of larger (macrocytic) or
            smaller (microcytic) cells; shape variations from the normal, e.g.,
            in poikilocytosis, may not be detected by automated analysis

In addition, there is the reticulocyte count (RET, RETIC, RETC), which is expressed as an absolute count or as a percentage of the RBC (Lowenstein, 1959; Evans and Fagg, 1994; Houwen, 1992). Measurements of ratios between reticulocyte subpopulations by fluorescent intensities (Fuchs and Eder, 1991a, 1991b; Andrews, 1995; Collingwood and Evans, 1995) or reticulocyte hemoglobin content are available, but not commonly used in toxicology studies with laboratory animal samples (Brugnara, 2003).

For further information on analytical methods, see Chapter 10.

ERYTHROCYTE MORPHOLOGY TERMINOLOGY

The cell morphology can be classified by size, shape, hemoglobin content, color, and inclusions. Some of terms used to describe red cell morphology and which you may encounter are given here:

Acanthocytes: Red cells with spicules of varied length that are irregularly distributed over the cell surface; these cells are often smaller and appear clublike.

Anisocytosis: Variable-sized erythrocytes.

Elliptocytes: Erythrocytes that appear elliptical, oval, or pencil or cigar shaped that can be normo- or hypochromic; in hemolytic anemias these cells may have spicules.

Hypochromasia: Erythrocytes usually having a pale central region that is about one-third of the total volume; if this area is proportionally increased, the cell is described as hypochromic. It is usually associated with inhibition of heme or globin synthesis, decreased MCH and MCHC values.

Macrocytosis: Presence of larger mature erythrocytes usually but not always accompanied by increased mean MCV and RDW values, i.e., not reticulocytes.

Microcytosis: Presence of smaller erythrocytes usually but not always accompanied by reduced MCV values.

Poikilocytosis: Variation in erythroid shape; this term is sometimes replaced by more specific cell morphology terminology.

Polychromasia: An alteration of the differential staining of reticulocytes and erythocytes with an increase of the bluish staining associated with RNA and the pinkish color of hemoglobin; this is seen in reticulocytosis and regenerative anemias.

Schistocytes: Fragmented or partial red cells that can be microcytic, or of irregular shapes. Often this term includes dacryocytes, which are tear- or pear-shaped cells with an extended tail. These cells may be oberved in intravascular hemolysis or following mechanical damage, e.g., during extracorporeal circulation in pharmacology studies.

Spherocytes: Red cells that have lost their biconcave shape with a decreased surface-area-to-volume ratio.

Stomatocytes: Cup-shaped cells that are smaller than normal and often develop into spherocytes; these cells are observed where there is an imbalance of circulating lipid levels.

Target cells: The cell membrane is increased in proportion to the cell volume, resembling a target or bull's-eye with a dark staining center surrounded by two rings of stained and unstained cytoplasm. These cells are observed where there is an imbalance of circulating lipid levels.

Terms used to describe erythrocytic inclusions are:

Adhesion of Red Blood Cells: Red blood cells may be observed to be clumped due to antibody coating the cell surface, and therefore are described as **agglutinated**, or the cells may appear to be stacked on the blood film (like a stack of coins), which is described as **Rouleaux** formation.

Basophilic stippling: Aggregated irregular or round darkly staining granules varying in size and number, associated with ribosomal clumps, mitochondria, and siderosomes.

Cabot rings: Threadlike complete ring inclusions that are remnants of spindle fibers.

Heinz bodies: Irregular-shaped precipitates of denatured hemoglobin near the cell periphery, and which may sometimes appear as blunt projections from the cell. These cells may be counted by microscopic examination after using supravital stains such as new methylene blue or brilliant cresyl green.

Howell-Jolly bodies: Round-shaped remnants of nuclear chromatin.

Pappenheimer bodies: Aggregates of ferritin, and the cell that, if stained positive with Prussian blue, is termed a siderocyte.

Given that erythrocytes are typically biconcave disks, measurements of diameters and mean cell volume are highly dependent on the method for measurement, as the erythrocyte shape is altered to enable measurements (see Chapter 10). In Table 3.1 the approximate relative mean values for adult animal red mean cell volumes are shown; these are freely adapted from several sources of published and unpublished data, and the data show that although the cell diameters are similar for many species, the mean cell volumes differ particularly for mice and cats, which have much lower MCV values.

TABLE 3.1

Approximate Erythrocyte Diameter, Mean Cell Volume, and Life Span in Several Laboratory Animals

Species	Approximate Erythrocyte Diameter (μm)	Approximate Mean Cell Volume (fl)	Life Span(days)
Mouse	5 to 7	44 to 48	38 to 50
Rat	6.2	50 to 62	45 to 68
Dog	5.5 to 7.5	66 to 77	100 to 110
Guinea pig	6.9 to 7.8	77 to 86	60 to 80
Rabbit	6.7 to 6.9	68 to 70	60 to 70
Hamster	5 to 7	55 to 70	60 to70
Cat	5.5. to 6.3	39 to 53	66 to 78
Nonhuman primate	6.5 to 7.5	50 to 90	86 to 105 (Rhesus monkeys)

ERYTHROKINETICS

The life span of erythrocytes is shorter in most laboratory animal species than in humans (approximately 120 days). Erythrocytic life spans can be measured using isotope techniques or cytometric techniques (Horky et al., 1978; Derelanko, 1987; Argent et al., 1993; Nusbaum, 1997), but an approximation of red cell life span can be calculated using the following formula (Vacha, 1983):

$$\text{Mean red cell life span (days)} = 68.9 \times (\text{body weight in kg})^{0.132}$$

As a general rule, the smaller the animal, the shorter the life span (see Table 3.1).

The reticulocyte life span is less than 2 days before it becomes a mature red blood cell in laboratory animals. In dogs, reticulocytes were observed to be released in a cyclic manner approximately every 14 days (Morley and Stohlman, 1969), but in this author's experience this periodicity is not obvious during safety assessment studies in dogs. The reticulocyte count reflects the number of circulating erythrocytes that are removed by the reticuloendothelial system and very minor blood losses, as the production and destruction of erythrocytes are normally balanced. Normally, the reticulocyte numbers will vary inversely to the red cell life span, so in order to maintain the number of circulating erythrocytes, the bone marrow turnover must increase if the red cell life span is shortened. This balance will not be maintained when the bone marrow erythropoiesis fails. When there are generalized toxic effects on the bone marrow, other cell lineages, i.e., neutrophils and then platelets, are affected before the erythrocytes.

Senescent red cells are removed by the macrophages of the reticuloendothelial system in the spleen, liver, and marrow. Hemoglobin is broken down to globin, and then to the constituent amino acids that reenter the amino acid pool for protein synthesis. Heme is degraded to urobilinogen and bilirubin, which are excreted in bile and urine (Figure 3.1). The removal of circulating erythrocytes includes some cells that are not senescent, and this process seems to be random. Immunological and nonimmunological pathways for the removal of erythrocytes have been identified. The immunological removal processes performed by the macrophages involve immunoglobulin G and complement. Nonimmunological mechanisms include peroxidation, altered antioxidant cell capacity, and alterations of the phopholipid cell membranes.

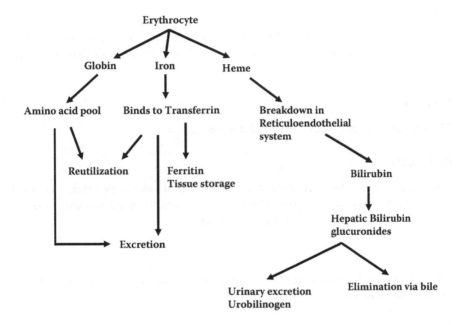

FIGURE 3.1   Red cell breakdown.

In Appendix A, a number of references are provided that give values for the various laboratory animal species, including some references that have collated data from various published sources. These references show some marked variations, and some reflect the analytical technologies at the time of measurement. In addition to the numbers, sizes, shapes, and life spans of erythrocytes, there are some morphological characteristics associated with the different laboratory animals, and these are briefly summarized next.

SPECIES CHARACTERISTICS

MOUSE

The erythrocytes show marked anisocytosis and polychromasia. Howell–Jolly bodies are frequently observed.

RAT

The erythrocytes show marked anisocytosis, with about 10% of cells being polychromatic.

BEAGLE DOG

Canine erythrocytes show mild anisocytosis and poikilocytosis. Nucleated red cells and Howell–Jolly bodies may be seen but are rare in blood films from healthy dogs. Reticulocyte counts are usually less than 3%.

RABBIT

The erythrocytes show marked anisocytosis, with up to 25% of erythrocytes being microcytic in some animals. Reticulocyte counts are usually less than 4% of red blood cells.

HAMSTER

The erythrocytes may show very mild polychromasia.

GUINEA PIG

The erythrocytes, which are larger than those of most laboratory species, are moderately anisocytotic with occasional microcytosis, and there are marked sex differences with lower erythrocytic values in females.

FERRET

The reticulocyte counts are reported to be variable from 2 to 14%, with Howell–Jolly bodies seen in about 5% of ferrets.

NONHUMAN PRIMATES

The published data show variations between Old World and New World primates and between animals that are captive bred versus wild caught.

MARMOSETS

These often show varying degrees of anisocytosis and polychromasia. The presence of Heinz bodies is not uncommon.

CAT

Feline erythrocytes with lower mean cell volume than other laboratory animals show moderate anisocytosis, and a tendency to Rouleaux formation. Macrocytic polychromatic red blood cells and reticulocyte counts are less than 0.5% of red blood cells. Heinz bodies are sometimes observed in cats.

ERYTHROCYTIC FUNCTIONS

The main role of red blood cells is to bind reversibly to oxygen using hemoglobin contained in the red cells. The structure of hemoglobin can by summarized as a tetrapyrrole ring system with central ferrous iron: there are four amino acid/globin chains per hemoglobin molecule, with two of the chains designated alpha chains and two beta chains. The chains are folded and carry a heme group attached to histidine residues at different positions on the alpha and beta chains. Heme synthesis involves porphyrins first by the condensation of glycine and succinyl coenzyme A to form delta levulinic acid (ALA), and a number of biochemical steps to form protoporphyrin. Iron is then combined with protoporphyrin to form heme.

The red cell membrane consists of fat (bipolar phospholipid), carbohydrate, and proteins. These membrane proteins are either integral and transmembrane, e.g., glycophorin or extrinsic and cytoskeletal (actin, ankyrin, and spectrin), and serve to maintain the red cell shape and deformability (Smith, 1987). The erythrocytic glycolytic pathway providing adenosine triphosphate (ATP) also maintains the red cell shape and deformability. The hexose monophosphate shunt pathway involving glucose-6-phosphate dehydrogenase provides a source of reduced nicotinamide adenine dinucleotide phosphate (NADPH), which maintains reduced glutathione (GSH) levels, and protects hemoglobin and the membrane protein from oxidant damage.

Among the stimulatory growth factors in erythropoeisis, the major promotor is the hormone erythropoietin (EPO), which is secreted by the renal interstitial cells in response to variations of circulating oxygen tension; so where the bone marrow is functional and hemoglobin is decreased, the consequences are lowered tissue oxygen tensions, with a stimulation of EPO production, which then results in an increase of erythrocyte production observed initially as reticulocytosis.

DERIVATIVES OF HEMOGLOBIN

The derivatives of hemoglobin may be found in blood and include the following:

Carboxyhemoglobin: Formed by combination of carbon monoxide.
Cyanmethemoglobin: Formed by the reaction of cyanide with methemoglobin, e.g., in cyanide poisoning.
Methemoglobin: Formed by oxidation of hemoglobin from the ferrous to the ferric state.
Sulfhemoglobin: Formed by severe oxidation and sometimes accompanies methemoglobin.
Sulfmethemoglobin: Formed by reaction of sulfide with methemoglobin.

Although the hemoglobins vary between and within species (Ingermann, 1997), these differences rarely present a problem when working with colonies of inbred laboratory animals.

Changes of Erythrocytic Measurements

There are two primary effects—reduction (anemia) or increase (polycythemia) of an individual or several red cell parameters.

Anemia

Anemia is defined as a decrease of hemoglobin, the number of circulating red cells, and alterations of mean corpuscular volume or mean corpuscular hemoglobin concentration, or it may be a combination of any of these measurements. In a broader view, anemia can be defined as the inability of blood to supply adequate oxygen for proper metabolic functions as a consequence of the erythrocytic changes. Where hemoglobin is decreased but the marrow remains functional, the reductions of tissue oxygenation stimulate erythropoietin production, and increased reticulocyte counts reflect this response.

Anemias result from either decreased production or increased destruction or loss of erythrocytes. Anemias also can be broadly divided into regenerative (responsive) or nonregenerative (nonresponsive) anemias based on the response or absence of bone marrow to a reduction in numbers of circulating red cells. Anemias can be further subdivided into those due to (1) blood loss, (2) increased destruction, (3) maturation defects, and (4) decreased production (Figure 3.2). There are several models used to classify anemias based on erythrocytic indices, duration, erythropoietic response by

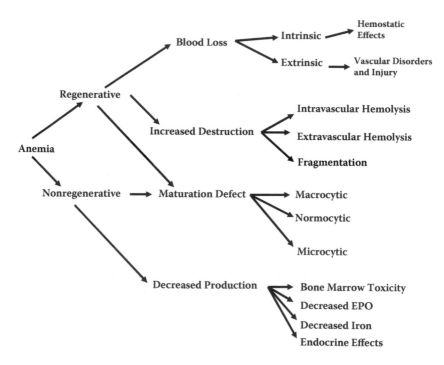

FIGURE 3.2 Broad classification of anemias.

the bone marrow, and the underlying mechanisms (Beutler, 1985; Hoffbrand, 1996; Tyler and Cowell, 1996; Fried, 1997).

The kinetics of the erythron are dependent on four distinct cell compartments: stem cells, progenitor cells, precursor cells, and mature erythrocytes. Perturbations of any of these compartments result in either anemia or erythrocytosis, and the intercompartmental kinetics vary between species. Complete inhibition of erythropoiesis in the mouse, rat, or dog has been estimated to result in 20, 10, or 5% reduction of red cell counts, respectively, assuming erythrocyte life spans of 25, 55, and 100 days for these species (Harvey, 1989). The reticulocyte numbers vary inversely to the red cell life span, so in order to maintain the number of circulating erythrocytes, the bone marrow turnover must increase if the red cell life span is shortened. Some indications of reticulocyte responses to blood loss are gained from studies where relatively large volumes of blood have been withdrawn (Berger, 1983; Capel-Edwards et al., 1981; Redondo et al., 1995). Erythrocyte changes can occur within 3 to 4 days, with reticulocyte responses observed within 5 to 10 days, and the number of reticulocytes sometimes dips before increasing.

MCV values can be used to classify anemias as microcytic (reduced) or macrocytic (increased cell volume), and the MCHC can be used to determine if the cells are hyperchromic (increased hemoglobin) or hypochromic (reduced). Anemias can be also be normocytic, e.g., in some aplastic anemias. In addition to the reticulocyte count, other measurements are also useful in defining the anemia; for example, RDW values may be used as an indication of the variation in blood cell volumes, and some additional measurements are discussed in Chapters 4 and 5.

Blood changes are dynamic, so a spectrum of findings may be seen over a time period; for example, during iron deficiency anemia erythrocytes may be normocytic and normochromic, and then at later stages the cells become microcytic and hypochromic. In short-term toxicity studies, the RBC may remain relatively unchanged but the reticulocyte count can be reduced, and this may be the first indication of altered bone marrow function. In a few cases, reticulocytopenia can result in an apparent lowering of the MCV due to reticulocytes being larger than erythrocytes, but this can be distinguished from microcytic anemia by examining the MCV values and cell distributions. The different erythrocyte life spans and bone marrow reserves of laboratory animals must be considered when interpreting the effects of xenobiotics, and in the design of studies, particularly those that involve a recovery phase.

Various anemias will be outlined in the following pages using the scheme outlined in Figure 3.2:

Blood loss
Increased destruction: Hemolytic anemias
   Oxidative anemia
   Immune anemia
   Red cell fragmentation
Maturation defect
   Macrocytic anemia
   Microcytic anemia

Decreased production
  Bone marrow toxicity
  Iron deficiency anemia
  Endocrine effects

Blood Loss Anemias

In toxicology studies, a common cause for these anemias is blood loss from the gastrointestinal tract, with compounds given orally and where compounds may cause irritation or ulceration. Another cause of significant blood loss may be due to poorly sited in-dwelling cannulae or surgical implant procedures. Blood loss can be acute, subacute, or chronic, with varied peripheral blood findings, dependent upon the amount of blood lost and the speed of the loss. Accelerated erythropoiesis with increased reticulocyte counts provides evidence of a response to the anemia, and these anemias generally show evidence of regeneration in surviving animals.

Interpretation may be complicated by the effects of fluid movements from the extravascular compartments that accompany blood loss. Chronic blood loss decreases the body's iron stores and may result in an iron deficiency; this deficiency stimulates a regenerative response, but where the blood loss is very severe and prolonged, this reduces erythropoiesis and the anemia becomes progressively less responsive. Blood loss may also occur as a consequence to vasculitis, coagulopathies, thrombocytopenias, fibrinolysis, and disseminated intravascular coagulopathies.

Increased Destruction

Red cell destruction occurs extravascularly in the macrophages of the reticuloendothelial system, bone marrow, liver, and spleen. Among the causes of intravascular hemolysis, which leads to increased plasma hemoglobin and hematuria, is hemolysis that may occur with intravenous injections when either the vehicle or xenobiotic alters the integrity of the red cell (see Chapter 4 for *in vitro* testing for hemolysis). Where there is also evidence of hemolysis in plasma or serum taken for biochemical measurements, it is important to rule out that the observed hemolysis is due to problems in either sample collection, subsequent sample processing, e.g., excessive centrifugation, contamination of plasma/serum with red cells at separation, or other sample processing problems.

Where hemoglobin breakdown occurs due to intravascular or extravascular hemolysis, it is possible to further identify the underlying mechanisms, and that these anemias are regenerative with the bone marrow compensating by increasing erythropoiesis. The inherited disorders of membrane and enzyme defects, e.g., glucose-6-phosphate dehydrogenase deficiencies and genetic defects of hemoglobins, play little part in preclinical toxicology unless compounds are designed deliberately to ameliorate the effects of these disorders; however, there may be serious consequences in human exposure to compounds that affect individuals where these deficiencies are present.

The mechanisms for acquired hemolytic anemias may be subdivided into those due to oxidation, immune responses, and red cell fragmentation, and those that are secondary to other disorders, e.g., hepatic and renal toxicity. The blood test results usually show low or low-normal hemoglobin, increased reticulocyte counts, and polychromasia, sometimes with altered erythrocyte shapes (sickle cells, spherocytes, elliptocytes). Plasma haptoglobin is reduced or not detectable, as increasing amounts of plasma hemoglobin are bound to the haptoglobin. The spleen weights and histopathology reveal splenomegaly and hematuria may be evident. The life span of erythrocytes in hemolytic anemias is reduced. Although radioactive isotope studies can be used to demonstrate reduced red cell life span, this is mainly of academic rather than practical use in preclinical drug development.

Oxidative Hemolysis

The inhibition of erythrocytic enzymes of the pentose–phosphate pathway and Embden–Meyerhof pathways can produce hemolysis with the oxidation of hemoglobin to methemoglobin, or sulfhemoglobin with or without the formation of Heinz bodies The oxidation may directly injure the cell wall by lipid peroxidation or change the hemoglobin molecule, which in some circumstances leads to its denaturation (Gordon-Smith, 1980; Edwards and Fuller, 1996).

Alternatively, some xenobiotics cause hemolysis by overcoming the glutathione redox-dependent cycle, which protects hemoglobin from oxidation in the erythrocyte, and included in the compounds that operate via the semiquinnone/glutathione redox interactions are aniline, anthracyclines, dapsone, naphthoquinnones, phenacetin, and menadione (Jenkins et al., 1972; Easley and Condon, 1984; Smith et al., 1985; Jensen and Jollow, 1991; Munday et al., 1991; Vage et al., 1994; Pauluhn, 2004, 2005). Some compounds are oxidized by hemoglobin, and it is the metabolites that then oxidize hemoglobin to methemoglobin, e.g., primaquine, phenylhydrazine (Brewer et al., 1962; Azen and Schilling, 1964; Bates and Winterbourn, 1984). The large number of oxidants producing methemoglobinemia include aminophenols and their derivatives, some antimalarials, chlorates, nitrites, hydroxylamines, sulfonamides, and aromatic compounds with amino nitro or hydroxyl groups. The increased production of methemoglobin and sulfhemoglobin is often associated with Heinz body formation.

Methemoglobin

Methemoglobin is also known as ferrihemoglobin due to the presence of ferric rather than ferrous heme iron. Methemoglobin is formed by endogenous oxidation and is usually present in low concentrations (<1%) in the blood due to its reduction primarily by methemoglobin reductase (Hjelm and de Verdier, 1965; Beutler, 1991).

In methemoglobinemias there are marked differences between species in their exposure to xenobiotics that cause oxidation of hemoglobin. These interspecies differences appear to be mainly due the ability of the erythrocytes to reduce the red cell methemoglobin concentrations. Rats and mice appear to have greater ability to reduce erythrocytic methemoglobin than dogs and humans, with nonhuman primates having a lower capacity (Marrs et al., 1987). If methemoglobinemia is observed in rodent studies, it is important to use a more suitable species such as the dog to assess the risk for humans: conversely very mild methemoglobinemia can be missed in rats because of their erythrocytic capacity for dealing with methemoglobin.

Sulfhemoglobin

The presence of sulfhemoglobin often accompanies the formation of methemoglobin, and its formation is not reversible, unlike methemoglobin.

Heinz Bodies

These inclusions found in red cells consist of denatured globin, and may be found in oxidative hemolytic anemias accompanying methemoglobin: occasionally this is termed Heinz body anemia. There are marked species differences in the production of Heinz bodies with the same test compound (Marrs et al., 1984). Of the various laboratory animal species, the marmoset and cat appear to be particularly prone to producing Heinz bodies.

Carboxyhemoglobin

Carbon monoxide (CO) binds to hemoglobin with an affinity that is more than 200 times that of oxygen to form carboxyhemoglobin, and this therefore reduces the capacity of hemoglobin to carry oxygen and inhibits the release of oxygen from the tissues.

Immune Hemolytic Anemia

Low-molecular-weight xenobiotics are not immunogenic, but they can form antibodies that then cause immune-mediated hemolysis by at least three mechanisms: (1) autoantibody induction, (2) formation of immune complexes, and (3) formation of haptens (Packman and Leddy, 1991). Warm antibody autoimmune hemolytic anemias typically involve immunoglobulin G showing maximum activity at body temperature, with microspherocytosis, polychromasia, anisocytosis, and occasionally nucleated red blood cells. The spleen is often enlarged, as the spleen plays an important role in the destruction of the antibody-coated red cells. In cold autoimmune hemolytic anemias, the antibody often is immunoglobulin M and shows maximum activity on cold storage at 4 degrees Celsius. The direct Coombs or antiglobulin test can be positive in drug-induced hemolytic anemias (see Chapter 4).

Some other examples include methyl dopa, which stimulates the formation of a "warm" type autoantibody, quinidine, which complexes with plasma protein to form an antibody that binds to the erythrocytic cell surface, and penicillin, which acts as a membrane-associated hapten. Some cephalosporins change the binding of the erythrocytic membrane with plasma immunoglobulins (Bloom et al., 1987). In these cases, a haptenic or immune complex anemia is sometimes found, but generally the red cell life span remains unchanged.

Red Cell Fragmentation

This fragmentation occurs in very severe anemias, in disseminated intravascular coagulation where small blood vessels have been damaged, and it may occur as a result of mechanical breakdown with some invasive procedures when blood is exposed to some in-dwelling catheterization or surgical procedures.

Maturation Defects

These anemias may be caused by alterations in the development of the erythropoietic cells of the marrow. These alterations may be in nuclear development or in the production of hemoglobin. Macrocytic anemia generally reflects abnormalities in nuclear development, whereas hypochromic microcytic anemia reflects abnormalities in hemoglobin synthesis.

Macrocytic Anemia

These anemias are characterized by larger erythrocytes—termed macrocytosis—with higher MCV values but reduced hemoglobin, with anisocytosis and poikilocytosis. The bone marrow morphology is altered with changes of erythroid, granulocytic, and megakaryocytic lineages. MCV may be increased in aplastic anemia, hepatotoxicity, pregnancy, and myelodysplastic syndromes, but the major type of macrocytic anemia is megaloblastic anemia, and these anemias are often caused by alterations of cobalamin (vitamin B_{12}) or folate metabolism.

In megaloblastic anemia due to cobalamin deficiency, the nuclear development of the erythroblasts is delayed and the nuclear chromatin is described as open and lacy; this is due to DNA synthesis. Xenobiotics may alter the absorption of metabolism of cobalamin or folate, or have a direct effect on DNA metabolism. In toxicology studies reductions of cobalamin are rare, partly because the vitamin content of animal diets is higher than the required dietary intake. Some xenobiotics can reduce vitamin B_{12} absorption intake, e.g., colchicine (Hoffbrand, 1996). Alterations of folate metabolism are more common than cobalamin deficiencies in toxicology, and effects include inhibition of dihydrofolate reductase, reduced folate absorption, and enzyme induction. See Chapter 5.

Microcytic Anemia

In microcytic anemia, the mean cell volume and hemoglobin are reduced. Usually these effects are due to alterations of heme synthesis that slow maturation processes. Iron deficiency anemia is sometimes classified as a maturation defect, but in this discussion it will be treated as a heme production-related anemia.

Decreased Production

Xenobiotics may affect hemoglobin production by toxic effects on the marrow cells, reducing heme synthesis, altering the rate of erythropoiesis via hormones or reduced renal EPO production. The inhibition of red cell production may or may not be accompanied by altered production of other cell lineages.

Bone Marrow Toxicity

The production of red cells by the bone marrow may be decreased (described as hypoplasia), completely inhibited (aplasia), abnormal (dysplasia), or malignant due to DNA alterations.

Aplastic Anemia

This is a result of bone marrow failure with a marked decrease of bone cellularity and pancytopenia. In pure red cell aplasia only the erythron is affected. Given the differing rates of production and turnover of the various cell populations, the effects on the blood picture vary, and the degrees of granulocytopenia and thrombocytopenia accompanying the anemia also vary, but in general, a pancytopenia is observed when the dose and exposure are sufficiently high. The degree of inhibition of hemopoiesis depends on the affected cell lines, the exposure to the xenobiotic or its metabolites, and duration, and varies with the animal species.

The erythroid lineage is particularly sensitive to xenobiotics that alter mitochondrial function, e.g., chloramphenicol succinate (Vincent, 1986; Turton et al., 1999; Ambekar et al., 2006). DNA synthesis may be inhibited by pyrimidine antagonists (e.g., hydroxyurea), 5-flurouracil, cytosine arabinoside, and purine antagonists (e.g., azathioprine, thioguanine).

Where the progenitor cells are affected, the bone marrow may not respond or may take some considerable time for the cells to recover. As a result of the effects on the leukocyte population, immunosuppression may also be observed with various cytotoxic drugs alongside the anemia. Rarely, some humans develop irreversible myelosuppression in a dose-dependent manner that appears to be related to the duration of dosing, and this sometimes occurs after the therapy has been discontinued. In some of these cases, it is postulated that there is some genotoxic injury in primitive cells that are then sensitized to the compound.

Reduced Heme Synthesis
Impairment of heme production may be due to abnormal porphyrin synthesis or a shortage of iron.

Porphyrins
There are several examples of compounds that disturb porphyrin metabolism and heme synthesis. Lead inhibits the ALA dehydratase and reduces ferrocheltalase activity; this results in increases of blood ALA, urinary copropophyrin III, and protoporphyrin (Feldman, 1999). Other porphyrinogenic chemical examples include the organopesticides lindane and heptachlor (Simon and Siklosi, 1974) and some herbicides. Isoniazid inhibits the formation of delta levulinic acid from pyridoxine. See Chapter 4.

Iron Deficiency
Mean cell volume and hemoglobin are reduced in severe iron deficiency anemia. Where there is excessive iron storage in the bone marrow this is described as *sideroblastic* anemia. In some erythroblasts iron granules are arranged in a ring around the nuclei of late erythroblasts, and these are termed ring sideroblasts. See Chapter 5.

Endocrine Effects
Erythropoeitin (EPO) is the major hormone governing erythropoiesis, and it is primarily produced by the kidneys, and to a lesser extent by the liver, although during the fetal period it appears to be produced mainly by the liver. The half-life of EPO is less than 6 hours, but it responds rapidly to hypoxia, and there is an inverse relationship between EPO and red cell mass under normal circumstances (Semenza, 1994).

In acute or chronic renal toxicity a normocytic anemia with reduced erythropoietin may be found. Rapid anoxia or severe blood loss may result in a quick release of immature reticulocytes. Where there is fluid loss, usually through the gastrointestinal tract, associated with body weight loss and diarrhea, and these fluid losses are prolonged, negative feedback on erythropoeitin production and erythropoiesis can occur. This can cause hypocellularity of the bone marrow and a reduction of the erythrocytic precursor cells. Anemias associated with the administration of androgens and estrogens have been reported (Finkelstein et al., 1944; Crafts 1946).

PSEUDOANEMIA

This may be caused by perturbations of fluid balance that lead to plasma expansion, thus causing apparent reductions of hemoglobin, red cell count, and hematocrit, i.e., hemodilution. In pregnancy the plasma volume may increase up to 50%, but the red cell mass does not increase to the same extent, resulting in a hemodilution effect. See Chapter 9.

RESPONSE TO AND RECOVERY FROM ANEMIA

Compared to other body cells, the blood cells have much shorter life spans, and these cells may be subjected to toxic effects both on the metabolism and function of mature circulating cells and on hemopoiesis. The number of circulating red cells is dependent on production and breakdown processes affecting these cells and it is more common in toxicity studies to observe cell number reductions caused by either impaired synthesis or increased breakdown than increases of red cell numbers. The cell life span needs to be considered when interpreting the effects of xenobiotics and in the planning of studies, particularly those that involve a recovery phase.

The reticulocyte numbers vary inversely to the red cell life span, so in order to maintain the number of circulating erythrocytes, normally the marrow turnover must increase if the red cell life span is shortened. Reticulocytosis occurs after a short period of hemopoietic suppression seen, for example, following a single large dose of some cytotoxic drugs. The hypoplastic event is followed by a hyperplastic corrective response and results in reticulocytosis. The reticulocyte count increases in number following acute hemorrhage and hemolytic anemias, but struggles to increase where there is a significant loss of developing erythroblasts such as in megaloblastic anemias. In shorter-term toxicity studies, the RBC may remain within the reference interval, but the reticulocyte count will be reduced, and this may be the first indication of altered bone marrow function, later to increase as erythropoietic response occurs. The reticulocyte absolute count is preferable to the percentage count for data interpretation. (In some cases of reticulocytopenia there is an apparent lowering of the MCV due to reticulocytes being larger than erythrocytes; this can be distinguished from microcytic anemia.) In chronic toxicity, red cell survival may be shortened and available iron reduced.

SECONDARY ANEMIA

There are several conditions that lead to anemia where the observed effects are secondary to other causes. Some examples are seen in hemolytic uremia syndrome, disseminated intravascular coagulation (DIC; see Chapter 8), endocrine disorders, and malnutrition (markedly reduced food consumptions). In severe hepatotoxicity, anemia, hemodilution, splenomegaly, and hemorrhage may occur with increased MCV, and the blood film examination shows the presence of target cells, echinocytes, and acanthocytes. Viral and bacterial infections may cause anemias, so in drug development using animal models of diseases, it is important to characterize the pretreatment baselines for infected and noninfected animals.

The majority of the heterogenous myelodysplastic syndromes are neoplastic disorders characterized by a pancytopenia and dysplasia in the presence of a normo- or hypercellular bone marrow, with the majority of these syndromes having chromosomal abnormalities. There are five subtypes: refractory anemia, refractory anemia with ring sideroblasts, refractory anemia with an excess of blasts, refractory anemia with an excess of blasts in transformation, and chronic myelomonocytic leukemia.

POLYCYTHEMIA

Absolute or true polycythemia describes conditions where the blood hemoglobin, red blood cell count, and total red cell mass are above normal levels. This is rare but may be caused by rapid stimulation of EPO production with increased erythropoiesis; cobalt appears to cause polycythemia via its effects on EPO production (Templeton, 1992).

Secondary polycythemia is associated with generalized tissue hypoxia as a result of hypoventilation or pulmonary damage.

Pseudopolycythemia, or relative polycythemia, is more common, and describes conditions where blood hemoglobin and erythrocytic parameters are elevated, but these changes are caused by reductions of plasma volume. Reductions of plasma volume occur in dehydration, following severe fluid loss from the gastrointestinal tract, stress, perturbations of fluid balance with accumulation of extracellular fluid, and with diuretics.

GENERAL COMMENTS

Xenobiotics may cause anemias by a number of mechanisms, including alterations of hemoglobin metabolism, red cell formation and breakdown, direct toxic effect on the red cell, and immunological or nutritional effects. As with many xenobiotic-induced toxic effects, there are marked differences between species when given the same compound. One example is with azido-3'-deoxythymidine (AZT), where erythroid hyperplasia is observed in rats but not mice (Luster et al., 1989; Thompson et al., 1991). Erythrocyte shapes have been reported when exposed to anionic amphipathic drugs to form echinocytes, and to form stomatocytes with cationic amphipathic drugs (Smith, 1987). Some drugs are designed to act on enzymes within the red cell, e.g., carbonic anhydrase inhibitors (Lin et al., 1991), and the potential consequences for the erythron must be considered when designing these studies and interpreting data. The therapeutic use of EPO and the availability of suitable assays have increased awareness of the roles played by EPO and potential changes in toxicity.

There are animal models and inherited diseases, particularly in dogs, which may be suitable for testing drugs aimed at ameliorating hereditary diseases.

Unless working with animal models of specific disease, it is difficult to detect any association between drug-induced anemias and inherited predispositions in general toxicology studies, for example, hemolytic anemia associated with glucose-6-phosphate dehydrogenase deficiency, where anemia will only be induced by a drug in the particular population with this enzyme deficiency. Comparative studies with known compounds that cause these effects may indicate a predisposition to cause similar effects of a novel compound.

In summary, alterations in erythrocytic parameters mainly but not always reflect an imbalance between production and loss of the erythrocytes. It is important to distinguish between regenerative and nonregenerative anemias. Direct injury to the hemopoietic stem cells will affect the production of erythrocytes, and in most cases the production of other cell lineages. Leukemias may also cause anemias because of bone marrow effects. Hepatotoxicity, renal toxicity, some endocrine toxicities, and severe inflammation can reduce erythropoiesis and cause anemias. Mild changes of red cell indices are not uncommon in toxicity studies, and these changes are often secondary to other toxicities. It is important to eliminate preanalytical variables as the causes of these changes. Some of the compounds causing anemias are listed in Appendix E.

REFERENCES

RETICULOCYTES

Andrews, C. M. 1995. The application for routine reticulocyte analysis to toxicology. *Sysmex J. Int.* 5:141–42.

Brugnara, C. 2003. Iron deficiency and erythropoiesis: New diagnostic approaches. *Clin. Chem.* 49:1573–78.

Collingwood, N. D., and Evans, G. O. 1995. Reticulocyte counts in laboratory animals using the R2000 flow cytometer. *Sysmex J. Int.* 5:126–29.

Evans, G. O., and Fagg, R. 1994. Reticulocyte counts in canine and rat blood made by flow cytometry. *J. Comp. Pathol.* 111:107–11.

Fuchs, A., and Eder, H. 1991a. Automated reticulocyte analysis of blood in different species. *Sysmex J. Int.* 1:34–38.

Fuchs, A., and Eder, H. 1991b. Zahl und reifegradverteilung der trikuolzyten von sechs tierarten. *Zentralbl.Veterinarmed A* 38:749–54.

Houwen, B. 1992. Review: Reticulocyte maturation. *Blood Cells* 18:167–79.

Lowenstein, L. M. 1959. The mammalian reticulocyte. *Int. Rev. Cytol.* 8:135–74.

LIFE SPAN OF ERYTHROCYTES

Argent, N. B., Liles, J., Rodham, D., Clayton, C. B., Wilkinson, R., and Baylis, P. H. 1993. A new method for measuring the blood volume of the rat using 113m indium as a tracer. *Lab. Anim.* 28:172–78.

Derelanko, M. J. 1987. Determination of erythrocyte life span in F344, Wistar and Sprague-Dawley rats using a modification of the [3H] diisopropylfluorophosphate ([3H] DFP) method. *Fundam. Appl. Toxicol.* 9:271–76.

Horky, J., Vacha, J., and Znojil, V. 1978. Comparison of life span of erythrocytes in some inbred strains of mouse using 14C-labelled glycine. *Physiologia Bohemoslovaca* 27:209–17.

Ingermann, R. L. 1997. Vertebrate hemoglobins. In *Handbook of physiology: Comparative physiology*, section 13, chap. 6. Vol. 1. New York: Oxford University Press.

Morley, A., and Stohlman, F. 1969. Erythropoiesis in the dog: The periodic nature of the steady state. *Science* 165:1025–27.
Nusbaum, N. J. 1997. Red cell age by flow cytometry. *Med. Hypotheses* 48:469–72.
Vacha, J. 1983. Red cell life span. In *Red blood cells of domestic mammals*, ed. N. S. Agar and P. G. Board, 67–132. Amsterdam: Elsevier.

GENERAL

Beutler, E. 1985. Chemical toxicity of the erythrocyte. In *Toxicology of the blood and bone marrow*, ed. R. D. Irons, 39–49. New York: Raven Press.
Fried, W. E. 1997. Evaluation of red cells and erythropoiesis. In *Comprehensive toxicology*, ed. J. C. Bloom, I. G. Sipes, C. A. McQueen, and A. J. Gandolfi, 35–54. Vol. 4. Oxford: Pergamon Press.
Hoffbrand, A. V. 1996. Megaloblastic anaemias and miscellaneous deficiency anaemias. In *Oxford textbook of medicine*, ed. D. J. Weatherall, J. G. G. Ledingham, and D. A. Warrell, 3484–500. Oxford: Oxford University Press.
Tyler, R. D., and Cowell, R. L. 1996. Classification and diagnosis of anaemia. *Comp. Haematol. Int.* 6:1–16.

BLOOD LOSS

Berger, J. 1983. The effects of repeated bleedings on bone marrow and blood morphology in adult laboratory rats. *Folia Haematol. Int. Mag. Klin. Morphol. Blutforsch.* 110:685–91.
Capel-Edwards, K., Wheeldon, J. M., and Mifsud, C. V. J. 1981. The effect of controlled daily blood loss on haemoglobin concentration, erythrocyte count and reticulocyte count of male rats. *Toxicol. Lett.* 8:29–32.
Harvey, J. W. 1989. Erythrocyte metabolism. In *Clinical chemistry of domestic animals*, ed. J. J. Kaneko, 196–97. San Diego: Academic Press.
Redondo, P. A., Alvarez, A. I., Diez, C., Fernadez-Rojo, F., and Prieto, J. G. 1995. Physiological response to experimentally induced anemia in rats: A comparative study. *Lab. Anim. Sci.* 45:578–83.

OXIDATIVE HEMOLYSIS

Azen, E. A., and Schilling, R. F. 1964. Extravascular destruction of acetylphenylhydrazine-damaged erythrocytes in the rat. *J. Lab. Clin. Med.* 63:122–36.
Bates, D. A., and Winterbourn, C. C. 1984. Haemoglobin denaturation, lipid peroxidation and haemolysis in phenylhydrazine-induced anemia. *Biochem. Biophys. Acta* 798:84–87.
Brewer, G. J., Tarlov, A. R., Kellermeyer, R. W., and Alving, A. S. 1962. The hemolytic effect of primaquine. XV. Role of methemoglobin. *J. Lab. Clin. Med.* 59:905–17.
Easley, J. L., and Condon, B. F. 1984. Phenacetin induced methemoglobinaemia and renal failure. *Anasthesiology* 41:99–100.
Edwards, C. J., and Fuller, J. 1996. Oxidative stress in erythrocytes. *Comp. Haematol. Int.* 6:24–31.
Gordon-Smith, E. C. 1980. Drug-induced oxidative haemolysis. *Clin. Haematol.* 9:557–86.
Jenkins, F. P., Robinson, J. A., Gellatly, J. B., and Salmond, G. W. A. 1972. The no-effect dose of aniline in human subjects and a comparison of aniline toxicity in man and the rat. *Food Cosmet. Toxicol.* 10:671–79.
Jensen, C. B., and Jollow, D. J. 1991. The role of N-hydroxyphenetidine in phenacetin-induced hemolytic anemia. *Toxicol. Appl. Pharmacol.* 111:1–12.

Munday, R., Smith, B. L., and Fowke, E. A. 1991. Haemolytic activity and nephrotoxicity of 2-hydroxy-1,4-naphthoquinone in rats. *J. Appl. Toxicol.* 11:85–90.

Pauluhn, J. 2004. Subacute inhalation toxicity of aniline in rats: Analysis of time dependence and concentration-dependence of hematotoxic and splenic effects. *Toxicol. Sci.* 81:198–215.

Pauluhn, J. 2005. Concentration-dependence of aniline induced methemoglobinemia in dogs: A derivation of an acute reference concentration. *Toxicology* 214:140–50.

Smith, M. T., Evans, C. G., Thor, H., and Orrenius, S. 1985. Quinone induced oxidative injury to cells and tissues. In *Oxidative stress*, ed. H. Sies. London: Academic Press.

Vage, C., Saab, N., Woster, P. M., and Svensson, C. K. 1994. Dapsone-induced hematologic toxicity: Comparison of the methemoglobin-forming ability of hydroxylamine metabolites of dapsone in rat and human blood. *Toxicol. Appl. Pharmacol.* 129:309–16.

METHEMOGLOBINEMIA

Beutler, E. 1991. Methemoglobinemia and sulfhemoglobinemia. In *William's hematology*, ed. W. J. Williams, E. Beutler, A. J. Erslev, and M. A. Lichtman, 743–45. 4th ed. New York: McGraw Hill.

Hjelm, M., and de Verdier, C.-H. 1965. Biochemical effects of aromatic amines. 1. Methemoglobinaemia, hemolysis and Heinz-body formation induced by 4,4'-diaminodiphenylsulphone. *Biochem. Pharmacol.* 14:119–28.

Marrs, T. C., Bright, J. F., and Woodman, A. C. 1987. Species differences in methaemoglobin production after addition of 4-dimethylaminophenol, a cyanide antidote to blood *in vitro*: A comparative study. *Comp. Biochem. Physiol.* 86B:141–48.

ENDOCRINE

Crafts, R. C. 1946. Effects of hypophysectomy, castration and testosterone propionate on hemopoiesis in adult male rat. *Endocrinology* 39:401–5.

Finkelstein, G., Gordon, A. S., and Charipper, H. A. 1944. Effect of sex hormones on anemia induced by hemorrhage in rat. *Endocrinology* 35:267–71.

Semenza, G. L. 1994. Regulation of erythropoietin production: New insights into molecular mechanisms of oxygen homeostasis. *Hematol. Oncol. Clin. N. Am.* 8:863–84.

IMMUNE HEMOLYTIC ANEMIAS

Bloom, J. C., Lewis, H. B., Sellers, T. S., and Deldar, A. 1987. The hematological effects of Cefonicid and Cefazedone in the dog: A potential model of cephalosporin hematotoxicity in man. *Toxicol. Appl. Pharmacol.* 90:135–42.

Packman, C. H., and Leddy, J. P. 1991. Drug related immunological injury of erythrocytes. In *William's hematology*, ed. W. J. Williams, E. Beutler, A. J. Erslev, and M. A. Lichtman, 681–87. 4th ed. New York: McGraw Hill.

APLASTIC ANEMIA

Ambekar, C. S., Lee, J., Kumana, C. R., and Liang, R. 2006. Succinate dehydrogenase: Possible molecular target for chloramphenicol succinate induced aplastic anemia and leukemia. *Clin. Chem.* 51(Suppl. A):89–90.

Turton, J. A., Yallop, D., Andrews, C. M., Fagg, R., York, M., and Williams, T. C. 1999. Haemotoxicity of choramphenicol succinate in the CD-1 mouse and Wistar Hanover rat. *Hum. Exp. Toxicol.* 18:566–76.

Vincent, P. C. 1986. Drug-induced aplastic anaemia and agranulocytosis, incidence and mechanism. *Drugs* 31:52–63.

OTHER REFERENCES

Feldman, R. G. 1999. *Occupational and environmental neurotoxicology.* Philadelphia: Lippincott-Raven.

Lin, J. H., Chen, I. W., and deLuna, F. A. 1991. Dose-dependent pharmacokinetics of MK-417, a potent carbonic anhydrase inhibitor, in experimental polycythemic and anemic rats. *Pharm. Res.* 8:608–14.

Luster, M. L., Germolec, D. B., White, K. L., Fuchs, B. A., Fort, M. M., Tomaszewski, J. E., Thompson, M., Blair, P. C., McCay, A., Munson, A. E., and Rosenthal, G. J. 1989. A comparison of three nucleoside analogs with anti-retroviral activity on immune and hematopoeitic functions in mice: In vitro toxicity to precursor cells and microstromal environment. *Toxicol. Appl. Pharmacol.* 101:328–39.

Marrs, T. C., Scawin, J., and Swanston, D. W. 1984. The acute intravenous and oral toxicity in mice, rats and guinea-pigs of 4-dimethylaminophenol (DMAP) and its effects on haematological variables. *Toxicology* 31:165–73.

Smith, J. E. 1987. Erythrocyte membrane structure, function and pathophysiology. *Vet. Pathol.* 24:471–76.

Simon, N., and Siklosi, C. S. 1974. Experimentelles Porphyrie-Modell mit Hexachlor-Cyklohexan. *Z. Hautkr.* 49:497–504.

Templeton, D. M. 1992. Cobalt. In *Hazardous materials toxicology*, ed. J. B. Sullivan and G. R. Krieger, 853–59. Baltimore: Williams and Wilkins.

Thompson, M. B., Dunnick, J. K., Sutphin, M. E., Giles, H. D., Irwin, R. D., and Prejean, J. D. 1991. Hematologic toxicity of AZT and ddC administered as single agents and in combination to rats and mice. *Fundam. Appl. Toxicol.* 17:159–76.

4 Additional Erythrocytic Measurements

In addition to the core tests for erythrocytes discussed in the previous chapter, there are several tests that can also be employed in studies to investigate toxicological mechanisms that cause anemias. Methods for the counting of reticulocytes and Heinz bodies and measurements of methemoglobin and sulfhemoglobin are included in Chapter 10. An *in vitro* hemolysis test should always be used for intravenous formulations prior to administration to animals. Many of these tests are often used occasionally or rarely, and they should be validated with compounds known to cause effects that are similar to those associated with the test compound. In this chapter, the following tests are considered:

Osmotic fragility
In vitro hemolysis
Plasma hemoglobin
Plasma haptoglobin
Hematuria
Fecal occult blood
Erythrocyte sedimentation rate
Antiglobulin test/Coombs test
Plasma lactate dehydrogenase
Glutathione
Oxygen dissociation curves
Porphyrins
Erythropoietin
Hemorrheology
Glycated hemoglobin
Micronucleus tests

The descriptions have been kept brief, as most of these tests are not commonly used in toxicology studies, but individually the tests may be useful in the elucidation of toxic mechanisms. Of these tests, only the test for hematuria is commonly used—partly because of its simplicity and combination with other parameters on a urinary test strip.

OSMOTIC FRAGILITY

The osmotic fragility of mammalian erythrocytes varies with the species, and a general though not absolute rule is that smaller erythrocytes are more susceptible to

43

lysis (Perk et al., 1964; Coldman et al., 1969; Hall and Follett, 1972). The biochemical characteristics of the erythryocyte membrane also play an important role in determining susceptibility to cell lysis (DeLoach et al., 1989). Osmotic fragility is determined by adding erythrocytes to a series of sodium chloride solutions of different osmolalities, and measuring the degree of hemolysis. This *in vitro* test is rarely performed in toxicology, although it can be useful in obtaining additional evidence of hemolytic effects and explaining interspecies differences. The use of an *in vitro* hemolysis assay should be a higher priority than the use of osmotic fragility tests.

IN VITRO HEMOLYSIS

Novel intravenous formulations should be tested for their suitability by incubating with whole blood and plasma, and measuring any hemolysis or protein precipitation that ensues. These studies should also be applied where any co-solvents (e.g., polyethylene glycols, propylene glycol, alcohol, or novel vehicles) are used to improve the solubility of the compound. Intravenous formulations should not cause significant levels of hemolysis, irritation, or protein precipitation *in vivo*.

In principle, the test is performed by adding blood at various dilutions to the intravenous formulation, and hemolysis is measured by measuring the hemoglobin in the supernatant. The intravenous formulation is also added to plasma at various dilutions, and the mixtures are observed for evidence of protein flocculation. Ideally, no hemolysis or protein flocculation should occur. Formulations that cause gross hemolysis or protein flocculation can be reformulated and retested prior to *in vivo* studies. Numerous methods have been described for performing these tests, but unfortunately there is no single recommended method or regulatory guideline for the assessment of *in vitro* hemolysis (Fievet et al., 1971; Prieur et al., 1973; Boelsterli et al., 1983; Reed and Yalkowsky, 1985, 1986, 1987; Obeng and Cadwallader, 1989; Salauze and Decouvelaere, 1994). *In vitro* techniques have been adapted by some investigators from a static test tube to a model that takes into account the pharmacodynamics associated with the injection rate of the compound and the blood flow rate at the site of injection (Dal Negro and Cristofori, 1996; Krzyzaniak et al., 1997). *In vitro* hemolysis methods have also been used to evaluate cytotoxicity (Isomaa et al., 1992), and the technique has potential when evaluating compounds for use in a susceptible human population such as individuals with glucose-6-phosphate dehydrogenase deficiency.

Several points for consideration in performing *in vitro* hemolysis studies emerge from the various published methods, and these points include:

> Species used for blood: Human blood is often used because *in vitro* formulations are generally targeted for use in human medicine, but an animal species may be more suitable for investigation where effects have been observed in preceding *in vivo* studies.
> Anticoagulant used to prevent blood clotting during the measurement.
> Number of individuals tested.

Ratios of blood to formulation: Often these are low ratios starting at 1:1, but some allowance must be made for the expected final distribution of the injected volume both locally when injected and in the total body blood volume.

Time and temperature used for incubating the blood with the formulation: If the period is much longer than anticipated for the distribution of the injected volume, slight hemolysis may occur that is unrelated to the test compound.

Methods for measuring hemoglobin should be sensitive, i.e., in the ranges for plasma hemoglobin, not whole blood.

Methods for protein precipitation may involve turbidometric measurements or visual observations.

Interpretation of the results obtained by these methods can be problematic, as static methods may yield positive *in vitro* results but with no adverse effects occurring in subsequent *in vivo* studies due to the distribution volume and rate of distribution.

PLASMA HEMOGLOBIN

Plasma hemoglobin may be measured by spectrophotometric methods or a dedicated bench-top analyzer using a method based on azidemethemoglobin formation (Vanzetti, 1966; Cripps, 1968; van Kampen and Zillstra, 1983). One of these methods should be applied in studies of *in vitro* hemolysis. A very simple check may be made on other samples taken at the same time to ensure that hemolysis is present in all samples from the same animal. If hemolysis is not present in all of these samples, then this suggests that it may be artifactual in origin. Measurement of free plasma hemoglobin is of very limited value in the diagnosis of chronic hemolytic anemias.

PLASMA HAPTOGLOBIN

In intravascular hemolysis, free hemoglobin is released into the plasma and bound to plasma haptoglobin, where the haptoglobin complex is then catabolized by the hepatic Kupffer cells. Haptoglobin may be reduced where there is acute or chronic hepatotoxicity, and where hemopoiesis is reduced. Haptoglobin is also a positive acute phase protein that may be elevated in inflammatory conditions, and this may occur with some hepatotoxins.

Several methods for measuring haptoglobin are based on the peroxidase activity of the hemoglobin-haptoglobin complex in mildly acidic conditions, or immunochemical methods using nephelometry or fluorometry (Solter et al., 1991; Walker et al., 1991; Katnik et al., 1998; Eckersall et al., 1999; Parra et al., 2005). Most methods require the use of calibration standards that are not species specific.

HEMATURIA

A reddish urine color may reflect the presence of blood, but the color may also be due to porphyrins or the test compound or metabolite. There are many commercially available test strips (dipsticks) for the detection of hemoglobin and red cells that are based on the pseudoperoxidase activity of hemoglobin, which is linked to color detection systems. The test strips give positive results with hemoglobin, myoglobin, and bacterial peroxidases; false negative results may result if the urines are not adequately mixed. Test strip results can be confirmed by microscopic examination of the urine sediments after low-speed centrifugation when erythrocytes may be observed (Cohen and Brown, 2003). Urines should be examined within a few hours of collection, while the cells remain intact for the confirmation of hemoglobinuria.

Blood may enter the urine from any point on the renal tract, and although some urologists have suggested that the source of hemoglobinuria, i.e., from pre- or post-glomerular sources, may be identified by the measurement of the urinary mean red cell volumes, this does not appear to work for laboratory animals where the urine osmolalities differ significantly from the ranges found in human urines.

FECAL OCCULT BLOOD

This test can be used to detect and confirm gastrointestinal bleeding and can be performed using guaiac detection systems based on the detection of the pseudoperoxidase activity of hemoglobin, or some newer tests using immunochemical detection methods. Although anti-human hemoglobin antibodies react with canine hemoglobin, assays should be checked to establish suitable cross-reactivities with laboratory animal blood. Methods using the detection of peroxidase activity are subject to interference from the presence of other peroxidases in the diet, and degradation of hemoglobin in the intestine, yielding false positive results (Cook et al., 1992; Rice and Ihle, 1994; Jinbo et al., 1998). Some compounds (or metabolites) may impart a red color to feces. As small amounts of blood may be naturally lost in the feces, it may be necessary to adjust the sensitivity of the tests for each species, as dietary peroxidase in feces may be higher in some species and cause false positive results. The sensitivity can be altered by changing the proportions of reagents (Dent, 1973). False positive results may be caused by contamination from blood losses, e.g., small animal paw cuts or open wounds. Techniques using radiolabeled isotopes for the measurement of blood loss from the gastrointestinal tract are rarely used in toxicology studies.

ERYTHROCYTE SEDIMENTATION RATE (ESR)

ESR measurements are generally not used in toxicology studies, partly because of their inherent variability and lack of diagnostic specificity, and partly because current technologies demand relatively high blood volumes to be taken. Centrifugal methods for determining ESR do not appear to be suitable for laboratory animals, as the detection of the plasma:cell interface is dependent on the measurement of plasma bilirubin levels, which are generally much lower in laboratory animals than in humans.

ANTIGLOBULIN OR COOMBS TEST

Immune hemolytic anemias may be confirmed by showing erythryocytic sensitization by immunoglobulin and complement components using a direct antiglobulin test

(Coombs test or DAT) or modification of the test, such as the direct enzyme-linked antiglobulin test (DELAT) (Coombs et al, 1945; Jones et al., 1990, 1992; Barker et al., 1992). Flow cytometry also offers a suitable alternative to the DAT tests where erythrocytes can be examined using appropriate antiglobulin sera (Garratty, 1990).

PLASMA LACTATE DEHYDROGENASE

This plasma enzyme may be elevated in some anemias (Van Lente et al., 1981; Kazmierczak et al., 1990), but this additional measurement is of little value in the diagnosis of anemia in laboratory animals given the high variability of total plasma lactate dehydrogenase (LDH) and the differing isoenzyme distribution in laboratory animals, particularly in rodents.

GLUTATHIONE

Glutathione (N-(N-L-gamma-glutamyl-L-cysteinyl) glycine) is a tripeptide thiol compound that maintains the ferrous state of iron in heme; glutathione also removes toxic peroxides in metabolic processes catalyzed by glutathione peroxidase (Beutler, 1984; Meister, 1994). The half-life of glutathione is about 3 to 4 days in dogs. Gluta-thione may be measured using commercially available enzyme-linked immunosor-bent assay (ELISA) techniques. Buthionine sufoximine is a useful model compound for inducing glutathione depletion (Khynriam and Prasad, 2001). The measurement should be considered if major alterations of glutathione occur with reduced red cell survival, and the measurement might help in mechanistic studies.

ERYTHROPOIETIN (EPO)

This glycoprotein with a molecular mass of approximately 34 kDa is an essential growth factor for hemopoiesis. It is synthesized mainly by the kidney and to a much lesser extent in the liver and other tissues. Plasma EPO is reduced in chronic renal injury and causes anemia. The secretion of EPO is stimulated by reductions of oxy-gen supply to the tissues, and the increased production of EPO in turn stimulates mixed cell lineages and red cell progenitor cells, causing pronormoblasts and early erythroblasts to proliferate (Aganostou et al., 1977; Okano et al., 1991; Krantz, 1991; Semenza, 1994). The half-life of EPO is less than 8 hours.

EPO can be measured by radioimmunoassay, ELISA, and chemiluminescence methods (Pechereau et al., 1997; Naeshiko et al., 1998; Marsden, 2006), but the available antibodies are not always suitable for all laboratory species (Wen et al., 1993). For some species, even when there is evidence of cross-reactivity with the antibodies, the assays need to adjusted for the lower ranges encountered.

OXYGEN DISSOCIATION CURVES

This test has very few applications in toxicology studies, but it may be used as an efficacy biomarker when testing compounds designed to shift the oxygen saturation curve in the treatment of sickle cell anemias (Bedell et al., 1984), or with the growing number of red cell substitutes being designed for human and animal transfusions.

PORPHYRINS

Some xenobiotics can cause porphyria, and some drugs may potentiate porphyria in susceptible human populations with hereditary conditions. Lead is the often-quoted example of a compound causing porphyria because it inhibits the amino laevulinic acid dehydratase (ALA) (Figure 4.1) in heme synthesis. Other examples include some organopesticides, e.g., lindane and herbicides (Taira and San Martin de Viale, 1998; Jinno et al., 1999).

Reddish-colored urines are produced in porphyria, and this finding must be distinguished from hematuria or other colorations due to test compound or metabolite. Diagnostic tests include erythrocytic protoporphyrin, plasma porphyrins, urinary porphyrins, urinary porphobilinogen, urinary delta-aminolevulinic acid, and fecal porphyrins (Deacon and Elder, 2001; Sandberg and Elder, 2004). Similarly, red coloration of feces can sometimes occur in perturbations of porphyrin metabolism, but more often the fecal color is due to either the test compound or blood. Urinary and fecal porphyrins may be extracted by alcohol or ether and examined for fluorescence under ultraviolet light. There are some more elegant techniques available using high-performance liquid chromatography.

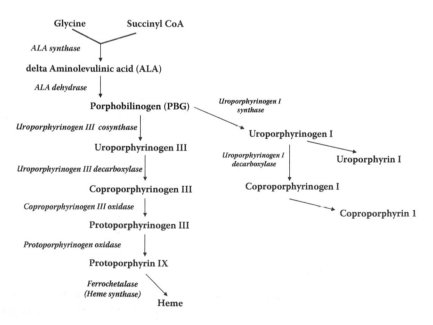

FIGURE 4.1   Porphyrin metabolism.

HEMORRHEOLOGY

The flow of blood depends upon the microcirculation—arterioles, capillaries, and venules under the arterial–venous pressure differentials—and the resistance of this network to blood flowing through it. Flow resistance also depends upon the deformability of individual and adherence of erythrocytes and leukocytes. Blood viscosity is related to a number of factors, including hematocrit, plasma fibrinogen, globulin, total leukocyte count, and red cell aggregatory properties.

Blood viscosity may be measured by several techniques, including capillary or rotational or porous-bed viscometers, but these techniques are rarely employed in regulatory toxicology studies (Dupont and Sirs, 1977; Pola et al., 1986; Somer and Meiselman, 1993), although they may have a role in safety pharmacology.

GLYCATED HEMOGLOBIN

Hemoglobins combine via amino groups with glucose and other sugars to form glycated hemoglobins, including HbA_{1a}, HbA_{1b}, and HbA_{1c}. *Glycation* is now preferred as a descriptive term replacing *glycosylation* or *glucosylation*. Erythrocytes are permeable to glucose, and a fraction of hemoglobin is glycated during the life span of erythrocytes, so glycated hemoglobin values will alter if the erythrocyte life span shortens or where significant blood loss occurs. Glycated hemoglobin measurements are being used increasingly to monitor long-term glucose homeostasis, particularly in the evaluation of antidiabetic agents.

There are now more than thirty different methodologies for measuring human glycated hemoglobins, and there continues to be a vigorous debate about international standardization and reporting unitage. Existing analytical methods include ion exchange chromatography, electrophoresis, affinity chromatography immunoassay, high-performance liquid chromatography, and enzymatic techniques. Not all methods are suitable for laboratory animals due to heterogeneity of the glycated hemoglobins in different species. Some ion exchange chromatography and immunochemical methods do not appear to give satisfactory results with animal samples, but affinity boronate chromatography appears to be suitable for rats (Nagisa et al., 2003; Evans, 2004, unpublished data). Glycated hemoglobins have been measured in mice, rats, dogs and monkeys (Dan et al., 1997; Higgins et al., 1982; Rendell et al., 1985; Kondo et al., 1989; Hasegawa et al., 1991; Yue et al., 1992; Cefalu et al., 1993; Hooghuis et al., 1994; Ukarapol et al., 2002).

MICRONUCLEUS TESTS

Although these tests are generally performed by genetic toxicology departments, and form part of the regulatory submission, there are opportunities for hematologists to contribute to the technologies and methods employed for micronucleus testing. The assay is employed to provide an *in vivo* method for the detection of clastogenic agents that interfere with mitotic cell division. Micronuclei are formed from displaced chromatin outside the main nucleus of dividing cells following telophase, and the micronuclei remain in the cytoplasm after the expulsion of nuclei. Increased numbers of micronuclei can be indicative of effects due to a genotoxic agent. Bone marrow and, more recently, blood are sampled at timed intervals (usually 24 and 48 hours) after the dosing of the test compound.

Manual counting of micronuclei is time consuming and observer dependent. The techniques are more difficult to apply to blood because of the low numbers of polychromatic erythrocytes. Various automated flow cytometric methods have been developed using various reagents, including thiazole orange, propidium iodide, and antibodies to CD71 transferrin receptor (Hayashi et al., 1984, 1992; Vanparys et al., 1992; Grawe et al., 1997; Hynes et al., 2002; Dertinger et al., 2003, 2004).

REFERENCES

Osmotic Fragility

Coldman, M. F., Gent, M., and Good, W. 1969. The osmotic fragility of mammalian erythrocytes in hypotonic solutions of sodium chloride. *Comp. Biochem. Physiol.* 31:605–9.

DeLoach, J. R., Gyongyossy-Issa, M. I. C., and Khatchtourians, G. G. 1989. Species-specific hemolysis of erythrocytes by T-2 toxin. *Toxicol. Appl. Pharm.* 97:107–12.

Hall, D. E., and Follett, A. J. 1972. Red cell osmotic fragility in the beagle dog: Normal values and diagnostic application. *Vet. Rec.* 91:263–66.

Perk, K., Frei, Y. F., and Herz, A. 1964. Osmotic fragility of red blood cells of young and mature domestic and laboratory animals. *Am. J. Vet. Res.* 25:1241–48.

In Vitro Hemolysis

Boelsterli, U. A., Shie, K. P., Brandle, E., and Zbinden, G. 1983. Toxicological screening models: Drug induced oxidative hemolysis. *Toxicol. Lett.* 15:153–58.

Dal Negro, G., and Cristofori, P. 1996. A new approach for evaluation of in vitro haemolytic potential of a solution of a new medicine. *Comp. Haematol. Int.* 6:35–41.

Fievet, C. J., Gigandet, M. P., and Ansel, H. C. 1971. Hemolysis of erythrocytes by primary pharmacologic agents. *Am. J. Hosp. Pharm.* 28:961–66.

Isomaa, B., Lilius, H., and West, A. 1992. Evaluation of cytotoxicity of the first MEIC chemicals by using haemolysis of human erythrocytes as an end point. *ATLA* 20:226–29.

Krzyzaniak, J. F., Alvarez Núñez, F. A., Raymond, D. M., and Yalkowsky, S. H. 1997. Lysis of human blood cells. 4. Comparison of *in vitro* and *in vivo* hemolysis data. *J. Pharm. Sci.* 86:1215–17.

Obeng, E. K., and Cadwallader, D. E. 1989. In vitro dynamic method for evaluating the haemolytic potential of intravenous solution. *J. Parenter. Sci. Technol.* 43:167–73.

Prieur, D. J., Young, D. M., Davis, R. D., Cooney, D. A., Homan, E. R., Dixon, R. L., and Guarino, A. M. 1973. Procedures for preclinical toxicologic evaluation of cancer chemotherapy agents: Protocols of the laboratory of toxicology. *Cancer Chemother. Rep.* 4:1–39.

Reed, K. W., and Yalkowsky, S. H. 1985. Lysis of human red blood cells in the presence of various cosolvents. *J. Parenter. Sci. Technol.* 39:64–69.

Reed, K. W., and Yalkowsky, S. H. 1986. Lysis of human red blood cells in the presence of various cosolvents. II. The effect of differing NaCl concentrations. *Parenter. Sci. Technol.* 40:88–94.

Reed, K. W., and Yalkowsky, S. H. 1987. Lysis of human red blood cells in the presence of various cosolvents. III. The relationship between haemolytic potential and structure. *J. Parenter. Sci. Technol.* 41:37–39.

Salauze, D., and Decouvelaere, D. 1994. In vitro assessment of the haemolytic potential of candidate drugs. *Comp. Haematol. Int.* 4:34–36.

PLASMA HEMOGLOBIN

Cripps, C. M. 1968. Rapid method for the estimation of plasma haemoglobin levels. *J. Clin. Pathol.* 21:110–12.
van Kampen, E. J., and Zillstra, W. G. 1983. Spectrophotometry of hemoglobin and hemoglobin derivatives. *Adv. Clin. Chem.* 23:199–257.
Vanzetti, G. 1966. An azide-methemoglobin method for hemoglobin determination in blood. *J. Lab. Clin. Med.* 67:116–26.

PLASMA HAPTOGLOBIN

Eckersall, P. D., Duthie, S., Safi, S., Moffatt, D., Horadagoda, N. U., Doyle, S., Parton, R., Bennett, D., and Fitzpatrick, J. L. 1999. An automated biochemical assay for haptoglobin: Prevention of interference from albumin. *Comp. Haematol. Int.* 9:117–24.
Katnik, I., Pupek, M., and Stefaniak, T. 1998. Cross reactivities among some mammalian haptoglobins studied by a monoclonal antibody. *Comp. Biochem. Physiol.* 119B:335–40.
Parra, M. D., Vaisanen, V., and Ceron, J. J. 2005. Development of a time-resolved fluorometry based immunoassay for the determination of canine haptoglobin in various body fluids. *Vet. Res.* 36:117–29.
Solter, P. F., Hoffman, W. E., Hungerford, L. L., Siegel, J. P., St. Denis, S. H., and Dorner, J. L. 1991. Haptoglobin and ceruloplasmin as determinant of inflammation in dogs. *Am. J. Vet. Res.* 52:1738–42.
Walker, A. K., Ganney, B. A., and Brown, G. 1991. Haptoglobin measurements in a number of laboratory animal species. *Comp. Haematol. Int.* 1:224–28.

HEMATURIA

Cohen, R. A., and Brown, R. S. 2003. Microscopic hematuria. *N. Engl. J. Med.* 348:2330–38.

FECAL OCCULT BLOOD

Cook, A. K., Gilson, S. D., Fischer, W. D., and Kass, P. H. 1990. Effect of diet on results obtained by use of commercial test kits for detection of occult blood in feces of dogs. *Am. J. Vet. Res.* 18:1749–51.
Dent, N. J. 1973. Occult blood detection in faeces of various animal species. *Lab. Pract.* 22:674–76.
Jinbo, T., Shimizu, M., Hayashi, S., Shida, T., Sakamoto, T., Kitao, S., and Yamamoto, S. 1998. Immunological determinations of faecal haemoglobin concentrations in dogs. *Vet. Res. Commun.* 22:193–201.
Rice, J. E., and Ihle, S. L. 1994. Effects of diet on fecal occult blood testing in healthy dogs. *Can. J. Vet. Res.* 58:134–37.

ANTIGLOBULIN OR COOMBS TEST

Barker, R. N., Gruffydd-Jones, T. J., Stokes, C. R., and Elson, C. J. 1992. Autoimmune haemolysis in the dog: Relationship between anaemia and levels of red blood cell bound immunoglobulins and complement measured by an enzyme-linked antiglobulin test. *Vet. Immunol. Immunopathol.* 34:1–20.

Coombs, R. R. A., Mourant, A. E., and Race, R. R. 1945. A new test for the detection of weak and incomplete Rh agglutinins. *Br. J. Exp. Pathol.* 26:255–66.

Garratty, G. 1990. Flow cytometry: Its application to immunohaematology. *Balliere's Clin. Haematol.* 3:267–87.

Jones, D. R. E., Gruffydd-Jones, T. J., Stokes, C. R., and Bourne, F. J. 1990. Investigation into factors influencing performance of the canine antiglobulin test. *Res. Vet. Sci.* 48:53–58.

Jones, D. R. E., Gruffydd-Jones, T. J., Stokes, C. R., and Bourne, F. J. 1992. Use of a direct enzyme-linked antiglobulin test for laboratory diagnosis of immune mediated hemolytic anemia in dogs. *Am. J. Vet. Res.* 5:457–65.

PLASMA LACTATE DEHYDROGENASE

Kazmierczak, S. C., Castelliani, W. J., Van Lente, F., Hodges, E. D., and Udis, B. 1990. Effect of reticulocytosis on lactate dehydrogenase isoenzyme distribution in serum: In vivo and in vitro studies. *Clin. Chem.* 36:1638–41.

Van Lente, F., Marchand, A., and Galen, R. S. 1981. Diagnosis of hemolytic anemia by electrophoresis of erythrocyte lactate dehydrogenase isoenzymes on cellulose acetate or agarose. *Clin. Chem.* 27:1453–55.

GLUTATHIONE

Beutler, E. 1984. *Red cell metabolism: A manual of biochemical methods.* 3rd ed. New York: Grune & Stratton.

Khynriam, D., and Prasad, S. B. 2001. Hematotoxicity and blood glutathione levels after cisplatin treatment of tumor-bearing mice. *Cell. Biol. Toxicol.* 17:357–70.

Meister, A. 1994. Glutathione–ascorbic acid antioxidant system in animals. *J. Biol. Chem.* 269:9397–400.

ERYTHROPOIETIN

Aganostou, A., Schade, S., Barone, J., and Fried, W. 1977. Effects of partial hepatectomy on extra renal production in anemic rats. *Blood* 50:457–61.

Krantz, S. B. 1991. Erythropoietin. *Blood* 77:419–34.

Marsden, J. T. 2006. Erythropoietin—Measurement and clinical applications. *Ann. Clin. Biochem.* 43:97–104.

Naeshiro, I., Yoshioko, M., Chatani, F., and Sato, S. 1998. Changes in the plasma erythropoietin level in rats following fasting, aging and anaemia. *Comp. Haematol. Int.* 8:87–93.

Okano, M., Ohnota, H., and Sasaki, R. 1991. Protein deficiency impairs erythropoiesis in rats by reducing serum erythropoietin concentrations and the population size of erythroid precursor cells. *J. Nutr.* 122:1376–83.

Pechereau, D., Martel, P., and Braun, J. P. 1997. Plasma erythropoietin concentrations in dogs and cats: Reference values and changes with anaemia and/or chronic renal failure. *Res. Vet. Sci.* 62:185–88.

Semenza, G. L. 1994. Regulation of erythropoietin production: New insights into molecular mechanisms of oxygen hemostasis. *Hematol. Oncol. Clin. N. Am.* 8:863–84.

Wen, D., Boissel, J. P., Tracy, T. E., Gruninger, R. H., Mulcahy, L. S., Czelusniak, J., Goodman, M., and Bunn, H. F. 1993. Erythropoietin structure-function relationships: High degree of sequence homology among mammals. *Blood* 82:1507–16.

Oxygen Dissociation Curves

Bedell, C. R., Goodford, P. J., Nein, J., White, R. D., Wilkinson, S., and Wootton, R. 1984. Substituted benzaldehydes designed to increase the oxygen affinity of human haemoglobin and inhibit the sickling of sickle erythrocytes. *Br. J. Pharmacol.* 82:397–407.

Porphyrins

Deacon, A. C., and Elder, G. H. 2001. ACP best practice no. 165. Front line tests for the investigation of suspected porphyria. *J. Clin. Pathol.* 54:500–7.

Jinno, H., Hatakayema, N., Hanioka, N., Yoda, R., Nishimura, T., and Ando, M. 1999. Cytotoxic and porphyrinogenic effects of diphenyl ethers in cultured rat hepatocytes: Chlonitrofen (CNP), CNP-amino, chomethoxyfen and bifenox. *Food Chem. Toxicol.* 37:69–74.

Sandberg, S., and Elder, G. H. 2004. Diagnosing acute porphyrias. *Clin. Chem.* 50:915–23.

Taira, M. C., and San Martin de Viale, L. C. 1998. Effect of lindane and heptachlor on delta aminolaevulinic acid synthase. *Arch. Toxicol.* 72:722–30.

Hemorrheology

Dupont, P., and Sirs, J. A. 1977. The relationship between plasma fibrinogen, erythrocyte flexibility and blood viscosity. *Thromb. Haemost.* 38:660–67.

Pola, P., Flore, R., and Tondi, P. 1986. Blood and plasma viscosity in experimentally induced hyper- and hypo-fibrinogenaemia. *Int. J. Tissue React.* 8:333.

Somer, T., and Meiselman, H. J. 1993. Disorders of blood viscosity. *Ann. Med.* 25:31–39.

Glycated Hemoglobin

Cefalu, W. T., Wagner, J. D., and Bell-Farrow, A. D. 1993. Role of glycated proteins in monitoring diabetes in Cynomolgus monkeys. *Lab. Anim. Sci.* 43:73–77.

Dan, K., Fujita, H., Seto, Y., and Kato, R. 1997. Relation between stable glycated hemoglobin A_{1c} and plasma glucose in diabetes-model mice. *Exp. Anim.* 46:135–40.

Hasegawa, S., Sako, T., Takemura, N., Koyama, H., and Motoyoshi, S. 1991. Glycated hemoglobin fractions in normal and diabetic dogs measured by high performance liquid chromatography. *J. Vet. Med. Sci.* 53:65–68.

Higgins, P. J., Garlick, R. L., and Bunn, H. F. 1982. Glycosylated haemoglobin in human and animal red cells. Role of glucose permeability. *Diabetes* 31:743–48.

Hooghuis, M., Rodriguez, M., and Castano, M. 1994. Ion exchange microchromatography and thiobarbituric acid colorimetry for the measurement of canine glycated haemoglobin. *Vet. Clin. Pathol.* 23:110–16.

Kondo, N., Shibayama, Y., Toyomaki, Y., Yamamoto, M., Ohara, H., Nakano, K., and Ienaga, K. 1989. Simple method for determination of A_{1c}-type glycated hemoglobin(s) in rats using high performance liquid chromatography. *J. Pharm. Methods* 21:211–21.

Nagisa, Y., Kato, K., Watanabe, K., Murakoshi, H., Odaka, H., Yoshikawa, K., and Sugiyama, Y. 2003. Changes in glycated haemoglobin levels in diabetic rats measured with an automated affinity HPLC. *Clin. Exp. Pharmacol. Phys.* 30:752–58.

Rendell, M., Stephen, P. M., Paulsen, R., Valentine, J. L., Rasbold, K., Hestorff, T., Eastberg, S., and Shint, D. C. 1985. An interspecies comparison of normal levels of glycosylated haemoglobin and glycosylated albumin. *Comp. Biochem. Physiol.* 81B:819–22.

Ukarapol, N., Begue, R. E., Hempe, J., Correa, H., Gomez, R., and Vargas, A. 2002. Association between *Helicobacter felis*-induced gastritis and elevated glycated haemoglobin levels in a mouse model of type 1 diabetes. *J. Inf. Dis.* 183:1463–67.

Yue, D. K., McLennan, S., Church, D. B., and Turtle, J. R. 1992. The measurement of glyco-sylated hemoglobin in man and animals by aminophenylboronic acid affinity chroma-tography. *Diabetes* 31:701–5.

Micronucleus Tests

Dertinger, S. D., et al. 2003. Micronucleated CD71-positive reticulocytes: A blood based end-point of cytogenetic damage in humans. *Mutat. Res. Gen. Toxicol. Environ. Mutagen* 542:77–87.

Dertinger, S. D., et al. 2004. Three-color labelling method for flow cytometric measure-ment of cytogenetic damage in rodent and human blood. *Environ. Mol. Mutagenesis* 44:427–35.

Grawe, J., Nüsse, M., and Adler, H.-D. 1997. Quantitative and qualitative studies of micronu-cleus induction in mouse erythrocytes using flow cytometry. I. Measurement of micro-nucleus induction in peripheral blood polychromatic erythrocytes by chemicals with known and suspected genotoxicity. *Mutagenesis* 12:1–8.

Hayashi, M., Norppa, H., Sofuni, T., and Ishidate, M. 1992. Mouse bone micronucleus test using flow cytometry. *Mutagenesis* 7:251–56.

Hayashi, M., Sofuni, T., and Ishidate, M. 1984. Kinetics of micronucleus formation in relation to chromosomal aberrations in mouse bone marrow. *Mutat. Res.* 127:129–37.

Hynes, G. M., Torous, D. K., Tometsko, C. R., Burlinson, B., and Gatehouse, D. G. 2002. The single laser flow cytometric micronucleus test: A time course study using colchicine and urethane in rat and mouse peripheral blood and acetaldehyde in rat peripheral blood. *Mutagenesis* 17:15–23.

Vanparys, P., Deknudt, G., Vermeiren, F., Sysmans, M., and Marsboom, R. 1992. Sampling times in micronucleus testing. *Mutat. Res.* 282:191–96.

5 Iron, Associated Proteins, and Vitamins

Iron-containing proteins play major roles as catalysts in oxidative processes (termed oxidative stress) that can affect DNA, lipids, and proteins, and where oxidation of these cellular components causes injury to the tissues. In chronic malignancies or chronic inflammatory disorders, tissue macrophages can be activated and bind with apotransferrin, which consequentially reduces the available plasma total iron-binding capacity and produces an iron deficiency anemia. Iron deficiency is the most common cause of anemias, and these iron-deficiency anemias often develop slowly. There are several drugs that have been associated with anemias caused by deficiencies of iron and its associated vitamins. There now is sufficient evidence together with an improved knowledge of the role of erythryopoietin and the availability of new measurements on some automated hematology analyzers (see "Potential Markers" section) to suggest this area of erythrocytic metabolism requires wider investigation, and therefore a separate chapter has been devoted to the topic.

IRON METABOLISM

Iron is present in hemoglobin, the reticuloendothelial system, muscle, plasma, and cellular enzymes, with more than half of the total body iron present as hemoglobin.

There are two stable states of iron that exist as either ferro (2+) or ferri (3+) forms, and iron plays an important part in numerous biochemical oxidations; perturbations of iron metabolism lead to specific toxicities, as many detoxication and cell respiration mechanisms are dependent on iron (Ryan and Aust, 1992; Papanikolaou and Pantopoulos, 2005). Iron is also important for myoglobin functions in muscle.

Iron distribution may be classified into three types:

1. Functional iron, e.g., hemoglobin, myoglobin, other heme proteins, cytochromes, catalase, and nonheme proteins such as xanthine oxidase
2. Storage iron, e.g., ferritin and hemosiderin, which are formed where iron is in excess of cellular requirements
3. Transport iron, which is bound to proteins in transferrin and lactoferrin

Dietary iron is solubilized by the stomach's acidic pH, before absorption in the duodenum. Ferrireductase reduces ferrous to ferric iron on the surface of the duodenal mucosa, and iron is then transported into the cell by a divalent metal transporter (DMT1), where iron can be stored in the mucosal cell as ferritin or be moved to the basolateral wall of the cell by mobilferrin and is then reoxidized by hephasetin. Hephasetin links with ferroportin (IREG/MTP1) to transport iron into the plasma, where it binds with circulating transferrin. Senescent red blood cells are degraded by macrophages, in the spleen, and iron is transported to and recycled in the bone marrow for hemopoiesis. Overall, iron balance is primarily maintained by dietary intake, with losses of iron via the gastrointestinal tract, urogenital system, and skin. Approximately 80% of iron metabolism occurs in the bone marrow via transferrin receptors, with less than 1% of body iron occurring as transport iron—mainly transferrin (Figure 5.1).

Several review articles are listed under the "General References" section of this chapter.

Iron deficiencies caused by dietary manipulations have been described in several species—mice, hamsters, rats, and dogs—with associated effects on erythrocytic measurements (Tavenor, 1970; Ranasinghe et al., 1983; Boyne and Arthur, 1990; Weeks et al., 1990; Rao and Jagadeesan, 1995; Redondo et al., 1995). Plasma iron levels are about twice as high in female rats than in male rats, although iron storage is lower in female rats (Tavenor, 1970). Diurnal variations of plasma iron levels occur in humans and several laboratory species of rats (Fox et al., 1974; Tarkowska, 1976).

TRANSFERRIN

Iron is carried by the transport protein plasma transferrin to the bone marrow or other iron storage depots. Within the bone marrow, transferrin laden with iron binds to transferrin receptors and iron is taken into the erythroblast by this receptor-mediated endocytosis. A proton pump causes the pH in endocytotic vesicles to decrease, and

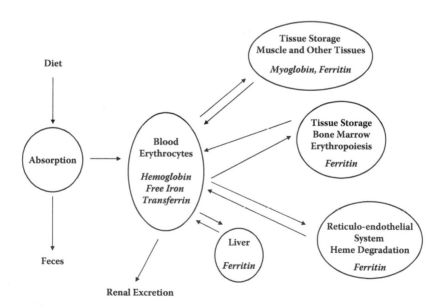

FIGURE 5.1 Iron distribution.

iron is then released from transferrin. Available iron is used for heme synthesis, with transferrin returning to the plasma for recirculation. Transferrin also transports iron from the reticuloendothelial system for recirculation after hemoglobin catabolism.

FERRITIN

This is the major iron storage protein, and ferritin is found mainly in the liver, spleen, and bone marrow. It has a molecular mass of about 470 kDa. Where iron exceeds the storage capacity of ferritin, the iron is then stored as hemosiderin. Although plasma ferritin is a useful measure of iron balance, the protein is also an acute phase protein increasing in inflammation. Both ferritin and transferrin synthesis are subject to changes caused by hepatic dysfunction, poor nutrition, renal disorders, and malignancies.

Plasma ferritin levels in rats are higher than in human serum (Zuyderhoudt et al., 1978), and female rats contain more hepatic ferritin than males (Bjørklid and Helgeland, 1970).

COBALAMIN (VITAMIN B$_{12}$)

Cobalamin is a water-soluble vitamin tightly bound to dietary protein. In the stomach pepsin and hydrochloric acid release cobalamin from dietary protein and bind it to R proteins. Pancreatic proteases release the R-bound cobalamin, which is then complexed with intrinsic factor, and this complex then adheres to specific receptors in the distal ileum. The uptake and transport of cobalamin involves three proteins: intrinsic factor, transcobalamin, and haptocorrin. Absorbed cobalamin binds to transcobalamin II (TCII), which carries cobalamin via the plasma to the liver, bone marrow, brain, and other tissues. Cobalamin is mainly attached to transcobalamin I in the plasma, but this is functionally inactive (Markle, 1996; Herbert, 1999).

Cobalamin (vitamin B$_{12}$) is a coenzyme for methionine synthase required for the demethylation of plasma 5-methyl tetrahydrofolate (methyl THF), which when demethylated enters cells to provide tetrahydrofolate (THF). THF in turn acts as substrate for the synthesis of intracellular folate polyglutamates, and the coenzymes required in pyrimidine and purine metabolism for DNA synthesis.

FOLATE

The coenzyme folate is required for the synthesis of thymidine monophosphate in the DNA synthesis pathway and folic acid (pteroylglutamic acid) is stored mainly in the liver. Folate is absorbed as monoglutamate following the action of intestinal folate conjugase, then reduced by dihydrofolate reductase to 5-methylhydrofolate, which is the main transportable form (Figure 5.2). There is a linkage between folate and cobalamin metabolism, which have synergistic effects, particularly where there is gastrointestinal toxicity (Branda, 1981; Bunch et al., 1990). Folate pools in rodents generally greatly exceed those in humans, so this species does not always show the effects of antifolate compounds.

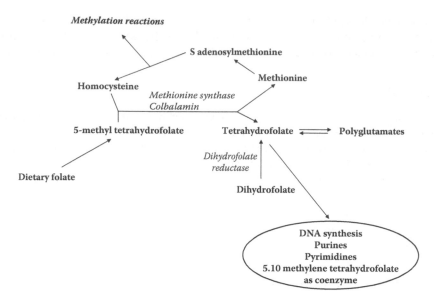

FIGURE 5.2   Folate metabolism.

ANALYTICAL METHODS

IRON

Plasma iron can be measured after treatment of a suitable reducing agent for the conversion of ferric to ferrous ions. The ferrous iron is complexed with ferrozine or other chromogenic reagents and measured colorimetrically.

Several references for laboratory animals are provided at the end of this chapter.

TRANSFERRIN AND FERRITIN

Although there are immunoassays for these proteins using nephelometry, immuno-turbidometry, or chemiluminescence, these have not been widely exploited for animal toxicology studies, primarily because there is often limited cross-reactivity with antibodies used for these measurements in human plasma. The use of ferritin assays is more challenging than the use of transferrin, as they require more sensitivity for the lower protein concentrations. However, measurements of ferritin and transferrin have been reported in several species.

Several references for laboratory animals are provided at the end of this chapter. Transferrin saturation can be expressed as percent, as it equals:

Plasma iron concentration × 100 when divided by plasma transferrin concentration, when both are concentrations expressed as µmol/l.

In the absence of a suitable assay for transferrin, the total or unsaturated iron-binding capacities can be measured and used more broadly across the laboratory species to assess plasma transferrin status.

Total Iron-Binding Capacity (TIBC)

The TIBC is the amount of iron required to saturate plasma transferrin. The measurement of TIBC involves sample pretreatment with the addition of an excess of iron solution to saturate the binding sites, and removal of the excess iron using one of several solid phase adsorbents; this is followed by measurement of the remaining iron. A direct method using pre- and post-pH adjustments is now available as an alternative. Although TIBC correlates with plasma transferrin, the relationship is not linear, so conversion factors should not be applied for animal samples.

Unsaturated Iron-Binding Capacity (UIBC)

By adding a known quantity of ferrous iron at an alkaline pH, and measuring the difference between the total amount added and the amount of unbound iron remaining, we get the UIBC value, and this represents the amount of iron bound by transferrin. The TIBC value is the sum of the plasma iron and UIBC.

Analyses of Cobalamin and Folate

The older microbiological assays for cyanocobalamin and folate were not convenient for use in toxicology studies, and these assays have been replaced by various immunoassays, and therefore are less tedious. Another advantage of these immunoassays is that they are not directly affected by drugs, unlike the microbiological methods where certain compounds, e.g., antibiotics and cytotoxics, could affect the growth of the microbiological organisms used for the analysis.

Potential Markers

For the future, soluble transferrin receptor and reticulocyte hemoglobin content are two measurements that appear to be sensitive to iron-deficient erythropoiesis, although further and thorough evaluation is required of these measurements in laboratory animals. These measurements may be made as stand-alone tests, but may have added value when used in conjunction with other measures of erythrocytes and reticulocytes, and then expressed as ratios (Beguin et al., 1988; Chitamber et al., 1991; Baynes et al., 1994; Punnonen et al., 1997; Brugnara, 2000; Cook et al., 2003; Hinzmann, 2003).

Soluble Transferrin Receptor (sTfR)

The soluble transferrin receptor (sometimes referred to as serum transferrin receptor) is a transmembrane molecule expressed on the surface of cells that require iron. Part of the molecule is shed as soluble transferrin receptor into the plasma. sTfR may be measured immunochemically by nephelometry or enzyme-linked immunoabsorbent assay (ELISA), but problems with the assays include cross-reactivities with animal samples and standardization.

ADDITIONAL RETICULOCYTE INDICES

With some automated blood analyzers, the hemoglobin content of reticulocytes can be measured (CHr), and it is calculated as the product of the hemoglobin concentration of single cells and the reticulocyte volume (MCVr) obtained by light-scattering properties.

Alternatively, the reticulocyte hemoglobin is expressed as an equivalent (RET-H) when obtained using fluorescence and light-scattering properties of reticulocytes. The fluorescent intensities may be used to identify immature reticulocytes.

sTfR, and CHr or RET-H appear to be promising sensitive indicators of iron deficiencies and early detectors of anemias in humans, but they require further evaluation in laboratory animals.

TOXIC EFFECTS

IRON

Effects on iron metabolism may be broadly categorized as iron deficiency, ineffective erythropoiesis, hemolysis, and iron overload, with the following features:

Iron deficiency (or chelation, blood loss) reduces plasma iron and ferritin, increases plasma transferrin (TIBC), and leads to microcytic and hypochromic anemias.

Ineffective erythropoiesis increases plasma iron and ferritin, and reduces plasma transferrin and reticulocyte counts.

Hemolysis increases plasma iron and ferritin, reduces transferrin, but increases reticulocyte counts.

Iron overload increases plasma iron and ferritin, reduces plasma transferrin, and leads to iron deposition in various organs and tissues.

Blood loss and gastrointestinal toxicity increase iron requirements due to increased loss of iron or reduced intake. Malnutrition with reduced protein calorie intake reduces plasma tranferrin.

Where total body iron is increased due to increased breakdown of red cells, iron is stored in the macrophages, followed by storage in the parenchymal cells. High levels of iron are associated with inflammatory actions involving granulocytes, and phagocytosis.

Iron given orally at high doses generates toxic oxygen metabolites with corrosive effects on the intestinal epithelium, and can cause oxidative damage to proteins, lipids, and DNA transitions. In sideroblastic anemias, the liver may show hemosiderosis and cirrhosis, and this may in turn affect hepatic biochemical markers. The ringed sideroblast may also be observed following the administration of isoniazid, lead, and in chronic alcohol models.

COBALAMIN AND FOLATE

Xenobiotics causing macrocytic anemias may be divided into those that act by:

1. Alterations of folate metabolism
2. Alterations of cobalamin metabolism
3. Directly acting on DNA metabolism

Blood films can be useful in the diagnosis of reduced cobalamin (vitamin B_{12}) or folate with the presence of macrocytes and hypersegmented neutrophils, and when more severe, to cause poikilocytosis and red cell fragments.

Malabsorption syndromes, changes of the bacterial flora of the gastrointestinal tract, and altered enterobiliary secretions can change the absorption of cobalamin and folate. Oral antibiotics and cytotoxics can alter the gut flora and reduce vitamin uptake. Loss of intrinsic factor causes cobalamin deficiency.

Compounds that interfere with intestinal folate uptake include phenylhydantoin, barbiturate, and alcohol.

Methotrexate competes with receptors for reduced folate and inhibits dihydrofolate reductase.

Cobalamin absorption may be decreased with histamine-2 receptor antagonists and proton pump inhibitors.

The effects on DNA may be direct on (1) pyrimidine synthesis, e.g., cytotoxic drugs such as 5-fluorouracil; (2) purine synthesis, e.g., 6-mercaptopurine; or (3) inhibitors of cellular enzymes, e.g., hydroxyurea.

REFERENCES

GENERAL REFERENCES

Andrews, N. C. 1999. Disorders of iron metabolism. *N. Engl. J. Med.* 341:1986–95
Andrews, N. C. 2000. Iron homeostasis: Insights from genetics and animal models. *Nat. Rev. Genet.* 1:208–17.
Brugnara, C. 2003. Iron deficiency and erythropoiesis: New diagnostic approaches. *Clin. Chem.* 49:1573–78.
Guyatt, G. H., Oxman, A. D., Ali, M., Willan, A., McIlroy, W., and Patterson, C. 1992. Laboratory diagnosis of iron-deficiency anemia: An overview. *J. Gen. Intern. Med.* 7:145–53.
Papanikolaou, G., and Pantopoulos, K. 2005. Iron metabolism and toxicity. *Toxicol. Appl. Pharmacol.* 202:199–211.
Ryan, T. P., and Aust, S. D. 1992. The role of iron in oxygen-mediated toxicities. *Crit. Rev. Toxicol.* 22:119–41.
Wick, M., Pinggerra, W., and Lehmann, P. 2000. *Iron metabolism, anemias, diagnosis and therapy.* 4th ed. New York: Springer-Wien.
Worwood, M. 1997. The laboratory assessment of iron status—An update. *Clin. Chim. Acta* 259:3–23.

DIETARY EFFECTS

Boyne, R., and Arthur, J. R. 1990. Anaemia and changes in erythrocyte morphology associated with copper and selenium dietary restrictions in rats. *Res. Vet. Sci.* 49:151–56.

Ranasinghe, A. W., Johnson, N. W., and Rountree, R. 1983. Experimental iron deficiency in Syrian hamsters, *Mesocricetus auratus. Lab. Anim.* 17:210–12.

Rao, J., and Jagadeesan, V. 1995. Development of a rat model for iron deficiency and toxicological studies: Comparison between Fischer 344, Wistar and Sprague-Dawley strains. *Lab. Anim. Sci.* 45:393–97.

Redondo, P. A., Alvarez, A. L., Dietz, C., Fernandez-Rojo, F., and Prieto, J. G. 1995. Physiological response to experimentally induced anemia in rats: A comparative study. *Lab. Anim. Sci.* 45:578–73.

Tavenor, W. D. 1970. *Nutrition and disease in experimental animals.* London: Balliere, Tindall and Cassell.

Weeks, B. R., Smith, J. E., and Stadler, C. K. 1990. Effect of dietary iron on hematologic and other measures of iron adequacy in dogs. *J. Am. Vet. Assoc.* 196:749–53.

Species—Iron

Bannon, P. D., and Friedell, G. H. 1972. Serum iron and iron binding capacity in normal and tumour-bearing golden hamsters. *Lab. Anim.* 6:75–78.

Fox, R. R., Laird, C. W., and Kirshenbaum, J. 1974. Effect of strain, sex and circadian rhythm on rabbit serum bilirubin and iron level. *Proc. Soc. Exp. Biol. Med.* 145:421–27.

Giulietti, M., La Torre, R., Pace, M., Iale, E., Patella, A., and Turillazzi, P. G. 1991. Reference blood values in cynomolgus macaques. *Lab. Anim. Sci.* 41:606–8.

Tarkowska, A. 1976. The diurnal variations of the serum iron level and the latent iron binding capacity in normal rats. *Ann. Univ. Marie Curie Sklodwoska (Med.)* 31:11–15.

Transferrin

Blumberg, B. S. 1960. Biochemical polymorphisms in animals. Haptoglobins and transferrins. *Proc. Soc. Exp. Biol. Med.* 104:25–28.

Palmour, R. M., and Sutton, H. E. 1971. Vertebrate transferrins. Molecular weights, chemical compositions and iron binding studies. *Biochemistry* 10:4026–32.

Regoeczi, E., and Hatton, W. C. 1980. Transferrin catabolism in mammalian species of different body sizes. *Am. J. Physiol.* 238:R306–10.

Rennie, J. S., MacDonald, D., and Douglas, T. A. 1981. Haemoglobin, serum iron and transferrin values of adult male Syrian hamsters (*Mesocricetus auratus*). *Lab. Anim.* 15:35–36.

Shifrine, M., and Stortmont, C. 1973. Hemoglobins, haptoglobins and transferrins in beagles. *Lab. Anim. Sci.* 23:704–6.

Ferritin

Andrews, G. A., Smith, J. E., Gray, M., Chavey, P. S., and Weeks, B. R. 1992. An improved ferritin assay for canine sera. *Vet. Clin. Pathol.* 21:57–60.

Bjørklid, E., and Helgeland, L. 1970. Sex difference in the ferritin content of rat liver. *Biochim. Biophys. Acta* 221:583–92.

Richter, G. W. 1967. Serological cross reactions of human, rat and horse ferritins. *Exp. Mol. Pathol.* 6:96–105.

Weeks, B. R., Smith, J. E., and Phillips, R. M. 1988. Enzyme-linked immunosorbent assay for canine serum ferritin, using monoclonal anti-canine ferritin immunoglobulin G. *Am. J. Vet. Res.* 49:1193–95.

Zuyderhoudt, F. M. J., Boers, W., Linthorst, C., Jorning, G. G. A., and Hengeveld, P. 1978. An enzyme-linked immunoassay for ferritin in human serum and rat plasma and the influence of the iron in serum ferritin on serum iron measurement, during acute hepatitis. *Clin. Chim. Acta* 88:37–44.

SERUM TRANSFERRIN RECEPTOR

Baynes, R. D., Skikne, B. S., and Cook, I. D. 1994. Circulating transferrin receptors and assessment of iron status. *J. Nutr. Biochem.* 5:322–30.
Beguin, Y., Hubers, H. A., Josephson, B., and Finch, C. A. 1988. Transferrin receptors in rat plasma. *Proc. Natl. Acad. Sci. U.S.A.* 85:637–40.
Chitamber, C. R., Loebel, A. L., and Noble, N. A. 1991. Shedding of transferrin receptor from rat reticulocyes during maturation in vitro: Soluble receptor is derived from receptor shed in vesicles. *Blood* 78:2444–50.
Cook, I. D., Flowers, C. H., and Skikne, B. S. 2003. The quantitative assessment of body iron. *Blood* 101:3359–63.
Punnonen, K., Irjala, K., and Rajamaki, A. 1997. Serum transferin receptor and its ratio to serum ferritin in the diagnosis of iron deficiency. *Blood* 89:1052–57.

RETICULOCYTE INDICES

Brugnara, C. 2000. Reticulocyte cellular indices: A new approach in the diagnosis of anemias and monitoring of erythropoietin function. *Crit. Rev. Clin. Lab. Sci.* 37:93–130.
Hinzmann, R. 2003. Iron metabolism, iron deficiency and anaemia. *Sysmex J. Int.* 13:65–74.

COBALAMIN (VITAMIN B$_{12}$)

Herbert, V., ed. 1999. Vitamin B12—An overview. In *Vitamin B12 deficiency*. Round Table Series 66. London: Royal Society of Medicine Press.
Markle, H. V. 1996. Cobalamin [Review]. *Crit. Rev. Clin. Lab. Sci.* 33:247–356.

FOLATE

Branda, R. F. 1981. Transport of 5-methyltetrahydrofolic acid in erythrocytes from various mammalian species. *J. Nutr.* 111:618–23.
Bunch, S. E., Easley, R. J., and Cullen, J. M. 1990. Hematologic values and plasma and tissue folate concentrations in dogs given phenytoin on a long-term basis. *Am. J. Vet. Res.* 51:1865–68.

6 Leukocytes

In this chapter, some of the leukocytic measurements, morphological comments, functions, and kinetics are discussed, followed by notes on various potential toxic effects. Measurements of the total and differential leukocyte counts are included in all studies requiring full blood counts, and the absolute rather than the percent counts should be used for interpretation. The roles of leukocytes in immunotoxicology are discussed separately in the following chapter. Preanalytical and analytical variables affecting leukocytic measurements are discussed in Chapters 9 and 10.

ABBREVIATIONS

The leukocytes are counted as five separate populations to obtain a five-part differential count. Although some analyzers produce a three-part differential count, this is not adequate for toxicological studies. Some abbreviations used with these five populations are:

Baso	Basophils
Eosn/eos	Eosinophils
Lymp/lym	Lymphocytes
Mono	Monocytes
Neut/neu	Neutrophils

However, there are no universally recognized single abbreviations except for WBC, indicating the total white blood cell count. Some laboratories use a single letter in reports to denote these counts, i.e., B, E, L, M, N. In some texts the neutrophil counts are subdivided into banded and segmented forms, but these counts have been largely bypassed with automated counters that provide the five-part differential count.

Lucs

These are large unstained/unclassified cells that are counted as part of the Bayer Diagnostics (Tarrytown, New York) analyzer technology. They are counted as a sixth component of the automated differential counts. These cells may include large activated lymphocytes, monocytes, and blast cells. See Chapter 10.

Granulocytes

Granulocytes are also known as the polymorphonuclear leukocytes, or polymorphs, with short, segmented nuclei and cytoplasmic granules. The term *granulocytes* is applied collectively to neutrophils, eosinophils, and basophils because of their granular contents.

LEUKOCYTE MORPHOLOGY TERMINOLOGY

The cell population numbers may be increased and termed *cytosis*, as in monocytosis, or *philia*, as in neutrophilia. Reductions of cell numbers are termed *penia* as in neutropenia. Some other terms used to describe leukocyte morphology and which you may encounter are given here:

Bands or stabs: Neutrophils with a nucleus that is U shaped or possesses lobes connected by a thick band rather than thread; less numerous than the segmented forms.

Döhle bodies: Large round or angular cytoplasmic inclusions in neutrophils and monocytes that result from the aggregation of the rough endoplasmic reticulum.

Erythrophagocytosis: Ingestion of one or several erythrocytes by neutrophils, monocytes, or macrophages.

Granulation: The number of staining characteristics of neutrophil granules varies in the different species. In toxic granulations associated with acute infections, the basophilias of the cytoplasm change in dogs and rats; this is in contrast to humans, where the granules become purplish and punctuate, indicating the early release of bone marrow neutrophils.

Hypersegmentation: Segmented neutrophils with more than five lobes to the nucleus and indicative of abnormal maturation.

Hyposegmentation: A lack of nuclear segmentation in the neutrophils.

Left and right shifts: Neutrophil maturity can be assessed by the development of nuclear lobes. Immature neutrophils are found to shift from the normal nuclear lobe number (e.g., three to five in dogs and man) to a nonlobed metamyelocyte stage, or in rodents as the presence of doughnut-shaped nuclei. These changes are described as a **left shift**. The neutrophil nuclear change or hypersegmentation that is seen in conjunction with megaloblastic development is described as a **right shift** when the average number of nuclear lobes is significantly increased.

Leukocytosis: This may involve one, several, or all of the different leukocyte populations.

Leukopenia: This may involve one, several, or all of the different leukocyte populations. Leukopenias may also occur as part of a pancytopenia.

Macrophage: A large mononuclear antigen presenting cell circulates in blood and tissues. These cells originate from the myelomonocytic stem cell and blood monocytes. When monocytes enter the tissues and body cavities, they differentiate into macrophages, which later may become activated and play vital roles in innate immunity, induction of immune tolerance, and tissue remodeling.

Pelger-Huët: Apparent left shift due to nuclei of granulocytes, particularly neutrophils, not maturing to segmented form.

Pycnosis or pycnotic: Describes a nucleus that is contracted into a strongly staining mass.

Segmented neutrophils: These develop from the banded forms.

Toxic cytoplasm: Diffuse basophilic cytoplasm in the neutrophils due to incomplete cellular maturation and utilization.

Vacuolation: Apparent spaces in the cytoplasm of neutrophils or lymphocytes in blood films, indicating the presence of vacuoles.

LEUKOCYTE KINETICS

A simplified description of leukopoiesis is given in Chapter 2, and this is shown again in Figure 6.1, with various cytokines acting as stimulators and inhibitors of the colony-forming units.

The life span of most of the leukocytic populations is much shorter than that of erythrocytes in the circulating blood (from a few hours to weeks), except for some lymphocytes that have a longer life span. However, the cells may remain in tissues for much longer periods.

Approximate cell life spans in the circulating blood are:

Neutrophils—About 8 hours
Lymphocytes—A longer life span than any other of the leukocyte major sub-
    populations; applies especially to some memory B cells
Monocytes—Circulate for about 25 hours
Basophils—Less than 6 hours
Eosinophils—Less than 1 hour

The life spans tend to be shorter in the smaller laboratory animals. Most laboratory animals show diurnal variations of leukocyte counts, so the timings of sample collections should be consistent or randomized when sampling from large groups.

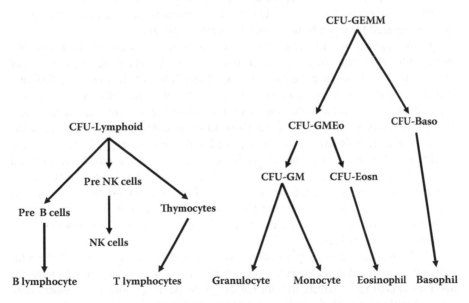

FIGURE 6.1 Development of leukocyte populations from colony-forming units.

For **neutrophils**, five distribution pools can be defined:

Proliferation (mitotic) pool
Maturation pool in the bone marrow
Marginal and circulating pools in blood
Tissue pool

In peripheral blood there are two pools—the circulating neutrophil pool and the marginal neutrophil pool—and neutrophils are interchanged between these pools. Neutrophils migrate through each of the pools to the tissues, where they remain for a few days before being destroyed. The neutrophils of the marginal pool are attached to the vascular epithelium, but when animals undergo strenuous exercise or become excited or stressed (mediated by epinephrine), neutrophils of the marginal pool are mobilized and produce a physiological neutrophilia. This pseudoneutrophilia can occur in toxicology studies.

Lymphocytes continually leave the blood via (1) endothelial venules in lymph nodes, Peyer's patches, and tonsils; (2) sinusoids in lymphoid organs; and (3) normal blood vessels in nonlymphoid organs. **Monocytes** also are located in a marginal pool in mice as well in the circulating pool. Although **basophils** spend a relatively short period in the circulating blood, they may survive for up to 14 days in the tissues. In rats, **eosinophils** complete their maturation in the spleen and spend only a short time in the circulation. So, although the majority of leukocytes have a relatively short life in the circulating blood, their life span in the tissues is much longer.

SPECIES CHARACTERISTICS

Blood cell counts vary with age, gender, etc., and they are subject to a number of pre-analytical factors (see Chapter 9). Here, the main characteristics of blood leukocytes are summarized for some healthy adult laboratory animals.

In the appendices, several references are given that contain microphotographs of blood to which the reader can refer, but this cannot replace the examination of blood films as part of the education process. These listed publications use different stains or microscope magnifications to demonstrate particular cell characteristics to the best advantage. In the local laboratory, the staining procedures and microscopic lens magnifications may differ from those in these publications. There are additional staining techniques that can be applied for the various cell types, but these are rarely used. The antigens expressed on the surface of a cell can now be recognized in the laboratory by using monoclonal antibodies, and this has led to the classification of these antigens, which is known as the cluster of differentiation (CD) system nomenclature system. Although the CD nomenclature is most frequently applied to lymphocytes (see Chapter 7), the nomenclature system now incorporates other hemopoietic cell markers.

Traditionally, leukocyte populations are expressed as percentages of the total leukocyte count, but it is important to recognize that in some instances the percentage may appear to be within the normal range, while the population numbers may be increased or decreased, so the absolute counts should always be considered. The variability in leukocyte populations counting is discussed in Chapter 10, and this relative inaccuracy must be considered, particularly when examining cell counts for basophils, monocytes, and eosinophils.

In the following paragraphs on the individual species, the intentions are to draw the reader's attention to the main characteristics and differences between species, e.g., the different neutrophil:lymphocyte ratios in dogs and rats, and the rarity of basophils in normal blood of most species in the laboratories.

In healthy animals, broadly the differential counts are as follows:

Mouse: Lymp >> Neut >> Mono, Eosn >> Baso
Rat: Lymp >> Neut >> Mono, Eosn >> Baso
Gerbil: Lymp >> Neut >> Eosn, Mono > Baso
Guinea pig: Lymp >> Neut >> Kurloff > Mono > Baso, Eosn
Rabbit: Lymp/Neut >> Mono, Baso, Eosn
Dog: Neut >> Lymp >> Mono, Eosn >> Baso
Nonhuman primate: Neut > Lymp >> Mono, Eosn, Baso

The proportions of monocytes, eosinophils, and basophils are much smaller than the proportions of lymphocytes or neutrophils. At a later age in the rat, the number of lymphocytes decreases, with a resulting reversal of the neutrophil:lymphocyte counts. Basophils are rare in most laboratory animals.

Dog

The blood neutrophil population is the most abundant leukocyte population, and it is about three- to fivefold the number of lymphocytes. Occasionally in a few dogs (Figure 6.2) the neutrophil:lymphocyte ratio may be reversed. Granules are almost absent in the neutrophil, but are faintly eosinophilic. Occasional band or immature neutrophils with elongated nuclei and with lighter-staining nuclear chromatin occur in low numbers. Lymphocytes are variable in size, with the smaller lymphocytes being more abundant, with densely clumped nuclear chromatin and scant amounts of cytoplasm. Some of the medium-sized lymphocytes are similar in size to the neutrophils, and they have a larger volume of cytoplasm surrounding the nuclei. "Reactive" lymphocytes are often larger with increased amounts basophilic cytoplasm.

Monocytes are larger than neutrophils with highly variable-sized and -shaped nuclei. The nuclear chromatin is less dense than the neutrophils, and there is a moderate amount of cytoplasm. Within the monocytes there are often cytoplasmic vacuoles, and occasionally some granules are found. Eosinophils are found in low numbers with variable numbers and sizes of prominent eosinophilic granules, and the nuclei are less lobulated than those of the neutrophils. Occasionally the eosinophil counts may be slightly increased and indicate subclinical infections, e.g., hookworm infestation. Basophils are rarely found in healthy dogs.

Rabbit

The neutrophil counts are similar to those for lymphocyte counts in mature animals, but the lymphocyte count is higher than neutrophils in young rabbits. The neutrophil

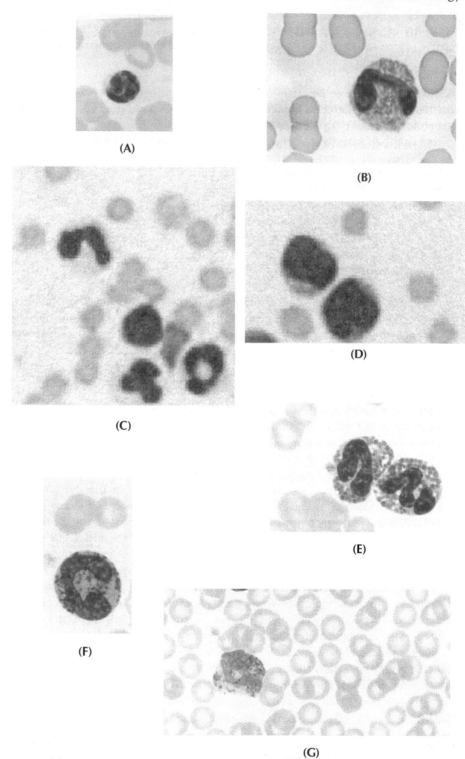

FIGURE 6.2 Dog: (A) lymphocyte, (B) neutrophil, (C) neutrophils, (D) monocyte, (E) eosinophil, (F) basophil, (G) basophil.

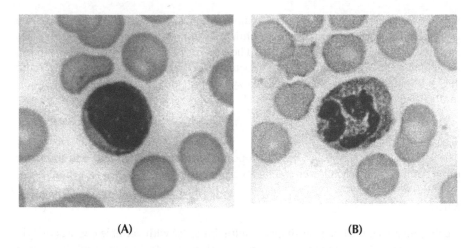

(A) (B)

FIGURE 6.3 Rabbit: (A) lymphocyte, (B) heterophil.

has a polymorphic nucleus surrounded by a variable number of granules, and therefore is sometimes described as a pseudoneutrophil or heterophil (Figure 6.3).

Lymphocytes of variable size have rounded pyknotic nuclei surrounded by a narrow band of cytoplasm. Monocytes are larger with amoeboid, lobulated, or horseshoe-shaped nuclei with cytoplasm that may be vacuolated. The eosinophils have bilobed or horseshoe-shaped nuclei, and these cells are larger than the heterophils with cytoplasmic acidic granules. Basophils contain metachromatic granules, and they are observed in small numbers. The Pegler–Huët effect is sometimes seen with nonsegmentation of leukocytes, mainly neutrophils.

GUINEA PIG

Lymphocytes are the dominant population of variable size with greater numbers of the smaller lymphocytes than the larger cells. The larger lymphocytes are similar in size to erythrocytes. The small lymphocytes have round pyknotic nuclei, whereas the nuclei are oval shaped in the larger lymphocytes, where cytoplasmic granules are found in the broader zone of the cytoplasm. The neutrophils are sometimes termed pseudoeosinophils, as they contain eosinophilic granules, and they have pygnotic and segmented nuclei. The larger monocytes have an oval, bean-shaped, or horseshoe-shaped nucleus. Basophils are rare, being similar in size to the neutrophils with lobulated or horseshoe-shaped nuclei and cytoplasmic granules. Unique to the guinea pig are the Kurloff cells, which may account for up to 4% of the total leukocyte count. This cell is a mononuclear leukocyte with intraplasmic mucopolysaccharide inclusions. These cells also are in the lungs, the red pulp of the spleen, thymus, and placenta; these cells are thought to act as killer cells.

HAMSTER

The hamster neutrophils also are referred to as heterophils, with lobulated pyknotic nuclei and round or rod-shaped granules. The lymphocyte sizes are variable, with smaller forms being numerous. The lymphocyte nuclei are also pyknotic, and the cytoplasm may contain granules. Basophil, eosinophil, and monocyte morphologies are similar to those of other rodents. There is a distinct diurnal variation with a leukocyte peak occurring during the nocturnal period when the animals are more active. Similar reported effects on leukocyte counts due to hibernation do not normally affect laboratory studies.

NONHUMAN PRIMATES

The data vary with the species, but the proportions of lymphocytes and neutrophils are often similar, with much lower counts for monocytes, basophils, and eosinophils.

MOUSE

Neutrophil nuclei may be ring shaped, multilobular, or with variable degrees of pyknosis and small discreet cytoplasmic granules (Figure 6.4). The neutrophils account for approximately 20 to 30% of the total leukocyte counts, with lymphocytes being more numerous and accounting for 70 to 80% of the total. The lymphocytes can be divided into two populations: small and large. The smaller lymphocytes are more numerous with rounded pyknotic nuclei. The nuclei of the larger lymphocytes have less pyknotic nuclei, sometimes with cytoplasmic granules. Eosinophils may account for up to 5% of the total leukocyte count, with annular or U-shaped nuclei and basophilic cytoplasm with acidophilic granules. Monocytes are the largest of the blood leukocyte types, with amoeboid or lobular nuclei surrounded by cytoplasmic granules and some degree of vacuolation. Basophils are rare, with large cytoplasmic granules.

RAT

Neutrophil nuclei are highly segmented, and there are fine granules in the neutrophil cytoplasm (Figure 6.5). The neutrophils account for between 10 and 38% of the total leukocyte count, with the lymphocytes being more numerous, accounting for 60 to 70% of the total. The lymphocytes are divided mainly into two populations: small and large. The smaller lymphocytes are more numerous and are approximately the size of the erythrocytes with cytoplasmic granules. The monocytes may account for 1 to 6% of the total leukocyte count, with bean-shaped or convoluted nuclei, and sometimes cytoplasmic granules. Eosinophils are slightly smaller than the neutrophils with densely packed, round cytoplasmic granules and an annular nucleus. Blood basophils are rare, with segmented or lobular nuclei, and they are smaller than the basophils found in tissues.

FUNCTIONS OF LEUKOCYTES

Primarily these cells act to protect the body from infection, working in conjunction with the various protein components—immunoglobulins and complement of the immune system—forming the host defense mechanisms. The leukocytes also produce cytokines that augment the function or stimulate proliferation of other cells.

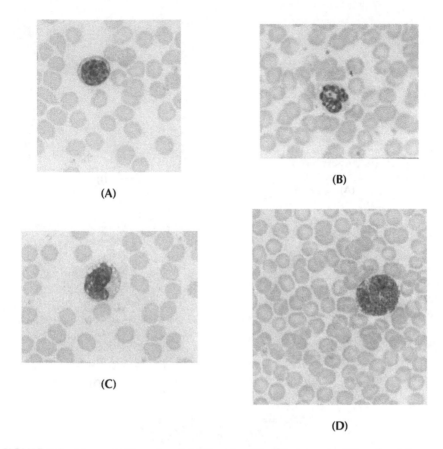

FIGURE 6.4   Mouse: (A) lymphocyte, (B) neutrophil, (C) monocyte, (D) eosinophil.

Neutrophils, eosinophils, basophils, and monocytes are all phagocytic—engulfing/ ingesting and destroying pathogens and cell debris. The phagocytes are attracted to bacteria at the site of inflammation (injection or inflammation at injection sites) by chemotactic substances released by injured tissues or complement. Phagocytosis is aided by the coating of cells or particles by immunoglobulin or complement, as phagocytes possess receptors for the Fc portion of immunoglobulin and complement C3b.

In contrast to erythrocytes and platelets, leukocytic actions are mainly extravascular, with the blood acting as a vehicle distributing the leukocytes to the tissues. Although the leukocyte population increases to counter bacterial and viral infections in untreated healthy laboratory animals, clinically significant infections are rare when animal care facilities and procedures are properly designed. Some minor infections can occur following small bites or where intravenous cannulae are not properly sited. The situation is different when the test compound is an immunostimulant or immunosuppressant.

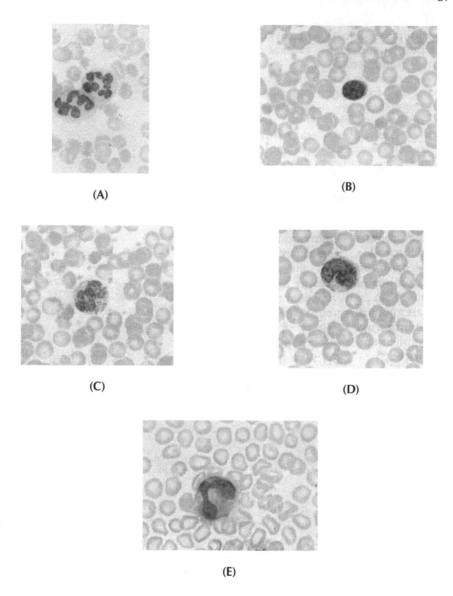

FIGURE 6.5 Rat: (A) neutrophil, (B) lymphocyte, (C) basophil, (D) monocyte, (E) eosinophil.

NEUTROPHILS

In response to chemotactic stimuli, neutrophil entry into tissues is aided by surface adhesion molecules that interact with the vascular epithelium. Neutrophil function tests are used to assess chemotaxis/migration and phagocytosis, degranulation, and the metabolic process involved in cell killing.

LYMPHOCYTES

These cells are produced in several sites, and blood contains only about 2% of all lymphocytes, with a short life span in blood. These cells are transported to various tissue sites, and lymphocytes continually leave the blood via different pathways: (1) the specialized endothelial venules in the T lymphocyte regions of the lymph nodes and Peyer's patches, (2) sinusoids in lymphoid organs such as the spleen, and (3) vessels of non-lymphoid tissues such as the lung, liver, and skin.

The mature lymphocytes are small mononuclear cells derived from hemopoietic stem cells. A common lymphoid stem cell undergoes differentiation and proliferation to give rise to two major subpopulations: B cells and T cells. B cell lymphocyte maturation occurs primarily in the bone marrow, and these cells mediate antibody immunity or humoral responses. T cells mature primarily in the thymus, and this also involves the lymph nodes, liver, spleen, and parts of the reticuloendothelial system (RES). T cells play a vital role in cell-mediated immunity. Both the receptor on T cells (TCR) and surface membrane immunoglobulin (sIg) on B cells are molecules that have constant and variable regions. Mature T cells can be subdivided into two main types: the helper cells expressing CD4 antigen and the suppressor cells expressing CD8 antigen. Natural killer (NK) cells are often large granular lymphocytes with prominent granules, and they are neither T nor B cells, although they may be CD8 positive. These NK cells can kill target cells by direct adhesion or bind to a target cell that has antibody bound to it (termed antibody-dependent mediated cytotoxicity [ADCC]). The bone marrow and the thymus are the primary lymphoid organs (see Chapter 7).

Monocytes enter the tissues from the circulation as macrophages, where they mature and carry out their principal functions, surviving for many days. They are mononuclear but have variable morphology; their cytoplasm is vacuolated, and they contain small granules with cytoplasmic projections often in the tissues. Monocytes are phagocytes that migrate primarily to the liver, spleen, alveoli, and peritoneum, where they develop into macrophages stimulated by inflammatory mediators. A small number of macrophages have the ability to proliferate. These macrophages perform diverse functions, removing senescent cells, and participate in antigen recognition, secreting a range of cytokines and modulating inflammatory responses.

The **monocyte–macrophage** system or reticuloendothelial system [RES] is formed by monocyte-derived cells and includes the Kupffer cells of the liver, alveolar macrophages in the lung, mesangial cells in the kidney, microglial cells in the brain, and macrophages within the bone marrow, spleen, lymph nodes, skin, and serosal surfaces. The spleen has important functions in filtering blood from the arteriolar circulation—the white pulp—through the endothelial mesh of the red pulp, and then to the venous sinuses. During these filtration processes, unwanted material from deformable red cells (hemosiderin granules, nuclear remnants) and particulate matters (e.g., opsonized bacteria) are removed.

The cells of the monocyte-macrophage system (RES) are localized particularly in tissues in which there is potential exposure to external pathogens and allergens, and the system enables cells to communicate with the lymphoid cells, and the liver, spleen, lymph nodes, bone marrow, thymus, and intestinal tract–associated lymphoid tissues. The functions of the monocyte-macrophage system include phago-cytosis and destruction of pathogens and cell debris, processing and presentation of antigens to lymphoid cells, and production of cytokines involved in the regulation of growth factor and cytokine networks governing hemopoiesis, inflammation, and cellular responses. The antigen presenting cells (APCs) react principally with T cells in the spleen, lymph nodes, thymus, bone marrow, and other tissues. An antigen-specific monocyte migration factor is produced by lymphocytes, and this causes monocytes to move and remain in areas where antigens are concentrated.

Eosinophils have kinetics and circulation similar to that of neutrophils, with IL-5 being an important factor regulating their production, and histamine as an important component in their granules. They are found mainly in the lung and gastrointestinal tract, apart from in blood (McEwen, 1992; Hirai et al., 1997).

Basophils are derived from the granulocyte precursor cells with characteristic large dark purple granules, which can obscure the nucleus. Their granules contain histamine and heparin, and immunoglobulin E (IgE) binding with surface receptors is involved in the release mechanisms. These cells are related to the small darkly staining cells in the bone marrow and tissues—the mast cells—which play an impor-tant role in defense mechanisms for parasites and allergens (Huntley, 1992; Metcalf et al., 1997).

TOXIC EFFECTS ON LEUKOCYTES

The cell population numbers may be increased and termed *cytosis* or *philia*; reduc-tions of cell numbers are termed *penia*. Xenobiotics may exert their effects on the leukocyte populations by suppression or destruction of the hemopoietic stem cells, the proliferating precursor cells, or the hemopoietic environment (see Chapter 2), or by inflammatory or microvascular changes. Reductions in the number of leukocytes may occur for more than one of the cell populations. These reductions may be part of a pancytopenia or be associated with primary anemias, or in some cases the leukope-nia is the dominant feature of a pancytopenia (Weiss, 1993). Reductions of cell num-bers for basophils, eosinophils, and monocytes are more difficult to detect because of the low numbers of cells in these populations under normal circumstances. It is very difficult to detect significant basophilic reductions when these cells are so rare in many laboratory species. Leukemias are a mixture of malignant diseases affecting the precursor cells, and these are briefly discussed under the heading "Neoplasias."

NEUTROPHILS

Neutrophilia may be the result of increased production by the bone marrow, increased release from the marrow granulocyte reserves, impaired exit from the peripheral blood, or a decrease of the proportion of neutrophils held in the marginal pool, with an associated increase in the circulating pool.

Causes of neutrophilia include bacterial infections, inflammation, trauma/sur-gery, stress, neoplasia, hemorrhage and hemolysis, myeloproliferative disorders, and pregnancy, and may be a secondary event, e.g., with neoplasms. Where these condi-tions are severe, the number of immature circulating neutrophils increases; this may be described as a left shift or a degenerative left shift, where the immature neutrophils outnumber the mature neutrophils. In hypoxia, there is a neutrophil migration from the marrow storage pool to both the marginal and circulating pools, and there may also be a shift from the marginal to the circulating pool. Chronic or acute administration of corticoids promotes neutrophil release from both the marrow and marginal pools; this effect is similar to the pseudoneutrophilia caused by exercise or stress. There may be shifts between the circulating and marginal pools without a change of the total number of neutrophils normally present when the numbers in these pools are combined.

Example compounds causing neutrophilia include steroids, insect venoms, lead, mercury, benzene derivatives, turpentine, and pyridine.

Neutropenia

There may be reduction of or ineffective proliferation starting in the mitotic pool of the bone marrow, leading to reductions of the other neutrophil pools. These effects may be on the stem cells, the colony-forming cells, or the bone marrow stroma. In severe aplastic anemia the effects on the stem cells cause neutropenias. Given the normally large reserve capacity of the bone marrow, effects on the bone marrow must be severe to cause neutropenia.

Alternatively, neutropenia may be caused by decreased production, increased consumption, or reductions of the marginal pool due to sequestration in the capillary beds. This can lead to increased production in the bone marrow.

Example compounds causing neutropenia include chloramphenicol, rifampicin, ristocetin, estrogens in dogs, phenylbutazone, phenothiazines, some sulfonamides, indomethacin, benzene, aflatoxin, cytostatic compounds, and procainamide. Irradiation causes similar effects.

Some compounds cause neutropenia via immune mechanisms where the compound binds onto the cell surface, leading to opsonization of cells, which are then attacked by bone marrow macrophages with a disruption of cell maturation processes. Some coated neutrophils may be released from the marrow and subsequently sequestered by the reticuloendothelial system. Immune-mediated neutropenias can by broadly divided into those that slowly develop and are often dose dependent, and those where the neutropenia develops more rapidly or neutropenia occurs after exposure to the compound ceases, sometimes many months afterwards.

Example compounds causing immune-mediated neutropenia include aminopyrine, chloramphenicol, and phenylbutazone.

Adhesion of granulocytes to the endothelium of blood vessel walls results in an increase of the marginal pool and a decrease of the circulating pool, and causes neutropenias.

Example compounds causing these effects are histamine, dextran, glucocorticoids, and iron salts.

Increased granulation is indicative of early release of neutrophils from the bone marrow pool—this is termed **toxic granulation**. This granulation effect varies greatly between species, being rare in the dog, where vacuolated and basophilic cytoplasm is more commonly seen in toxicity affecting neutrophils.

It is not solely changes in the number of cells that are caused by xenobiotics—the function of neutrophils, e.g., in phagocytosis, may also be altered, and although important, the **neutrophil function tests** have yet to find a regular place in toxicology studies. These are briefly discussed in Chapter 8.

LYMPHOCYTES

In the development of compounds aimed at immunostimulation or immunosuppression, the lymphocytes are key hematological efficacy biomarkers. Lymphocytic disorders can involve (1) reaction to infections, (2) functional disorders of the marrow-derived (B) cells or thymus-derived (T) cells, or both, and (3) malignancy. These disorders may be due to effects on the bone marrow, lymph nodes, or thymus, or be related to alterations in lymphocyte function.

Lymphocytosis

Lymphocyte counts may be increased in chronic infections, macrocytic anemia, lymphocytic malignancy, inflammation, and thyrotoxicosis, and there may be an increase of reactive or atypical lymphocytes, with deeper cytoplasmic basophilia and immature nuclei. Physiological lymphocytosis may also be observed following excitement or stress.

Lymphocytopenia

This often accompanies neutropenias. Reductions in the number of lymphocytes may be accompanied by altered or impaired lymphocyte functions. Lymphocyte subsets can be defined by the expression of their surface molecules and their end effects in immune responses relating to the release of soluble cytokines. Lymphocyte counts may also be reduced in moribund animals.

Example compounds causing lymphocytopenia include cytostatic compounds, compounds causing aplastic anemias, and glucocorticoids.

MONOCYTES

Monocytosis

This may occur in protozoal or bacterial infections, malignancy, and chronic inflammation. Some chronic myeloid leukemias in rodents are accompanied by evident partial monocytic differentiation. In laboratory animals, monocytosis often accompanies neutrophilia, and it may also occur following accidental injury or lung dosing of rodents in oral gavage studies.

Example compounds causing monocytosis include steroids at high dosages, glucocorticoids in dogs, and chlorpromazine.

Monocytopenia

Monocytopenias are difficult to detect given the low numbers of circulating cells, but can occur following severe bone marrow toxicity.

Example compounds include glucocorticoids in mice and compounds producing severe marrow toxicity.

BASOPHILS

Basophilia

Some causes of basophilia include parasitic infections, chronic inflammation, hematological malignancy, some leukemias, and chronic hemolytic anemias.

Basopenia

Given the low numbers of circulating cells, basopenias are difficult to detect, but can occur following severe bone marrow toxicity.

Example compounds include adrenocorticotrophin, estrogens, and antithyroid drugs.

EOSINOPHILS

Eosinophilia

Some causes of eosinophilias include parasitic infections, connective tissue disorders, hematological malignancy, and some leukemias. Increased eosinophil numbers have been reported in the gastrointestinal tracts of animals receiving modified sugars and starches (sorbitol, mannitol, lactose, and polyethylene glycol). The presence of increased numbers of eosinophils in the tissues, particularly in the lung and gastrointestinal tract, with irritant compounds may cause eosinophilia.

Example compounds include para-aminosalicylcic acid, some sulfonamides, gold salts, and methotrexate.

Eosinopenia

Given the low numbers of circulating cells, eosinopenias are difficult to detect but can occur following severe bone marrow toxicity. Corticosteroids and catecholamines have been implicated in eosinopenia.

NEOPLASIAS

In chronic carcinogenicity studies, hematologists may be asked to examine blood films and provide additional information on the myeloproliferative disorders found by histopathologic examination. More emphasis is placed on the pathology findings than the hematologic findings. The examination of bone marrows may help to confirm affected cell lineages. It must noted that leukemias can develop in control as well as dosed animals in rodent carcinogenicity studies. The effects due to the same xenobiotic may differ in the species used for carcinogenicity testing, with effects in one but not both species. The total leukocyte count may be also be markedly increased, not increased, or appear to be reduced to hemodilution in animals with neoplasms.

Hematological malignancies probably arise from a single cell in the bone marrow, thymus, or peripheral lymphoid system, and this single cell mutates, leading to a malignant transformation. Further mitotic division produces a clone of cells, and some of these cells may mutate further to produce subclones—a process termed clonal evolution. These subclones are transformed, can be resistant to apoptosis (programmed cell death), or proliferate excessively with either inappropriate activation or expression of the oncogene. The activation of the oncogene may be caused by amplification, point mutation, or translocation from one chromosomal location to another. Less commonly, the neoplasia may be due to action on the antioncogene or tumor suppressor genes that encode the important proteins that act to suppress cell growth. The antioncogene may be altered on one chromosomal allele, while mutation or deletion of the remaining allele may lead to uncontrolled cell growth.

Severe damage to chromosomes may be recognized by the genetic toxicologist as aneuploidy, chromosomal aberrations, formation of micronuclei, chromatid exchange with other chromosomes, and abnormalities in cell cycle kinetics.

Example compounds causing tumors include alkylating drugs, benzene and some organochemicals, chlorophenol, ethylene oxide, phenoxy acids, asbestos, styrene, polystyrene, vinyl chloride, heavy metals, and some cytostatic agents.

Classification of tumors almost always provokes debate among pathologists, and there are no universally accepted criteria (Greaves and Fancini, 1984; Ward et al., 1990; Stromberg, 1992; Raskin, 1996). Table 6.1 is a very simplified overview.

LEUKEMIAS

Most leukemias arise in the bone marrow and are characterized by high numbers of circulating white cells. Leukemias are a heterologous group of malignant diseases involving the precursors of peripheral blood cells. Leukemias follow genetic damage, e.g., caused by ionizing irradiation or exposure to chemical clastogens. The DNA repair mechanism may not be sufficient following the alteration, and this results in irreversible damage. In some instances a secondary insult overcomes the natural DNA repair mechanisms, and abnormal cell clones proliferate. Genetic toxicology plays a very important role in identifying the potential of xenobiotics to produce these chromosomal changes.

Acute Leukemia

Here, the number of hemopoietic blast cells increases to constitute more than 30% of the bone marrow, with the blast cells and primitive cell forms being found in both the peripheral blood and tissues. Dependent on the progression of leukemia, they are classified as acute or chronic. In acute leukemias primitive and blast cells are found in the blood and bone marrow. The malignant cells show either a chromosomal translocation or other DNA mutations affecting the oncogenes and antioncogenes. The acute leukemias can be subdivided into acute lymphoblastic leukemia (ALL)

TABLE 6.1
Simplified Overview of Malignant Blood Diseases

1. Leukemias: Malignant disorders of circulating blood cells originating from bone marrow

    Myeloid:

            Acute myeloid leukemia (AML)

            Acute monocytic leukemia (AMoL)

            Chronic myeloid leukemia (CML)

            Erythroleukemia

    Lymphatic:

            Acute lymphatic leukemia (ALL)

            Chronic lymphatic leukemia (CLL)

2. Lymphoma: Malignant disorders of cells found in lymph nodes affecting plasma cells, giant
    cells, or other cells of the lymph nodes

and acute myeloid or myeloblastic leukemia (AML); these can be subdivided based on morphological, immunophenotyping, and cytogenetic criteria.

Chronic Leukemia

More mature cells in varied development stages are found in chronic leukemias. The proliferation of leukemic cells may have a repressive effect on normal hemopoiesis, which results in anemia, and acute leukemia is additionally characterized by anemia, thrombocytopenia, and sometimes neutrophilia. Coagulation may be abnormal, and disseminated intravascular coagulation may be a feature.

Chronic Myeloid Leukemia (CML)

This is a clonal myeloproliferative disorder characterized by neutrophilia and neutrophil precursors—myelocytes and metamyelocytes. The basophils may be increased and the bone marrow is hypercellular with an increased number of granulocyte precursor cells and a raised myeloid:erythroid cell ratio. Some chronic myeloid leukemias in rodents are accompanied by evident partial monocytic differentiation.

MYELODYSPLASIA (MDS)

This clonal disorder is characterized by a peripheral blood cytopenia where there is macrocytic anemia and neutropenia. The neutrophil may be hypogranular with bilobed nuclei, termed pseudo-Pelger cells. The bone marrow is often hypercellular but may be hypocellular or fibrotic. MDS may be further subdivided into refractory anemia, which may present with neutropenia and thrombocytopenia, and refractory anemia where ring sideroblasts are found, and these may constitute more than 15% of the total erythroblasts.

OTHER TESTS APPLIED TO LEUKOCYTES

LEUKOCYTE ALKALINE PHOSPHATASE

Alkaline phosphatase can be demonstrated in the banded and segmented neutrophils contained in fresh blood and bone marrow smears using cytochemical staining techniques, and some compounds can increase the leukocytic phosphatase.

Example compounds include oral contraceptives and cortisol. The enzyme activity is low in chronic myelocytic leukemia.

MYELOPEROXIDASE

Myeloperoxidase is a hemoprotein found in neutrophils and secreted during their activation. This enzyme plays an important role in neutrophil microbial action, and meyloperoxidases have been implicated in cardiovascular disease, inflammation, the pathogenesis of Alzheimer's syndrome, and other neurological disorders. Analyzers using peroxidase staining techniques to count leukocytes can provide a mean peroxidase index (MPXI), which is indicative of the peroxidase activity of neutrophils.

PHOSPHOLIPIDOSIS

Many cationic amphiphilic compounds (CADs) cause the accumulation of abnormal lamellar lysosomes and crystalloid inclusions in tissues, including the lung, liver, kidney, lymph nodes, and several other tissues. Usually pathology findings are sufficient for the confirmation of phospholipidosis, but blood measurements may be used to confirm or demonstrate progression and recovery with phospholipidotic compounds. Nile red can be used as a vital stain for demonstrating the presence of these phospholipids in monocytes, and with flow cytometry, this offers a method for monitoring drug-induced phospholipidosis (Xia et al., 1997).

REFERENCES

Greaves, P., and Fancini, J. M. 1984. *Rat histopathology. A glossary for use in toxicity and carcinogenicity studies.* Amsterdam: Elsevier.

Hirai, K., Miyamasu, M., Takaishi, T., and Morita, Y. 1997. Regulation of the function of eosinophils and basophils. *Crit. Rev. Immunol.* 197:325–52.

Huntley, J. F. 1992. Mast cells and basophils: A review of their heterogeneity and function. *J. Comp. Pathol.* 107:349–72.

McEwen, B. J. 1992. Eosinophils: A review. *Vet. Res. Commun.* 16:11–44.

Metcalf, D. D., Baram, D., and Mekori, Y. A. 1997. Mast cells. *Physiol. Rev.* 77:1033–79.

Raskin, R. E. 1996. Myelopoiesis and myeloproliferative disorders. *Vet. Clin. N. Am. Small Anim. Pract.* 26:1023–42.

Stromberg, P. C. 1992. Changes in the hematologic system. In *Pathobiology of the aging rat,* ed. U. Mohr, D. L. Dungworth, and C. C. Capen, 15–24. Washington, DC: ILSI Press.

Ward, J. M., Rehm, S., and Reynolds, C. W. 1990. Tumours of the haematopoeitic system. In *Pathology of tumours in laboratory animals*, ed. V. Turosev and U. Mohr, 625–45. Vol. 1. Lyon, France: International Agency for Research on Cancer.

Weiss, D. J. 1993. Leukocyte response to toxic injury. *Toxicol. Pathol.* 21:135–40.

Xia, Z., Appelkvist, E.-L., DePierre, J. W., and Nassberger, L. 1997. Tricyclic antidepressant-induced lipidosis in human monocytes *in vitro* as well as in a monocyte derived cell line, as monitored by spectrofluorimetry and flow cytometry after staining with Nile Red. *Biochem. Pharmacol.* 53:1521–32.

7 Immunotoxicology

This chapter provides an introduction to immunotoxicological assays as hematologists have the opportunity, knowledge, and experience to make some contributions to this scientific topic. There are natural bridges between the disciplines of hematotoxicology and immunotoxicology in terms of the basic science, cells of common interest, cellular mechanisms, and people skills associated with cytometry. Chapters 2 and 6 introduced the functions and possible toxic effects on the main leukocyte populations; the effects on the immune system are broader, and additional tests are required to detect toxic effects on the immune system.

Several assays for investigating immunotoxicity are described, but there continues to be a debate about which tests should be used and when (see some of the numerous papers concerning strategies for immunotoxicity testing). There is no doubt that this science will evolve in the next decade given the increasing use of vaccines and biological agents, and a clearer picture will emerge with assays that are sufficiently robust for wider use.

COMPONENTS OF THE IMMUNE SYSTEM

In Chapter 2, the development of lymphocytes during hemopoiesis was traced, and this cellular development continues. The lymphocytes and other cells of the immune system originate from a network of organs including the bone marrow, thymus, spleen, and lymph nodes. The primary cellular components are the T and B cells, null cells including the natural killer (NK) cells, monocytes/macrophages, antigen presenting cells (APCs), dermal Langerhans cells, polymorphonuclear neutrophils (PMNs), and plasma cells. In addition, there are soluble mediators—the cytokines, immunoglobulins and complement proteins. The function of the immune system is to protect the host animal against viruses, bacteria, "foreign" cells and substances, and neoplasms (Figure 7.1).

Cells generated from the lymphoid stem cells are processed via the thymus and become T lymphocytes (T cells); other cells develop into B lymphocytes (B cells). The natural killer (NK) cells develop from the bone marrow.

T AND B LYMPHOCYTES

The lymphocytes can be subdivided into two main populations (see Chapter 6), with T cells developed in the thymus and B cells developed in the bursal tissues of the bone marrow, lymph nodes, and tissues of the gastrointestinal trac including Peyer's patches. These cells express surface membrane antigens (cluster differentiation [CD] antigens; see Appendix F), and these can be recognized by an ever-growing list of specific

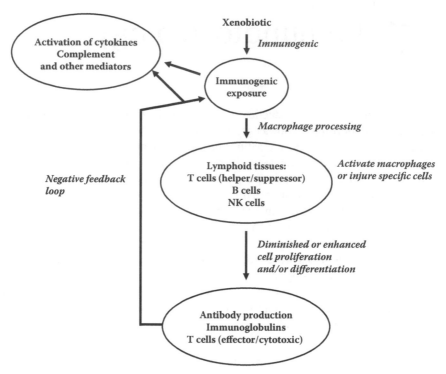

FIGURE 7.1.   Simplified immune system indicating potential effects of xenobiotics.

monoclonal antibodies. These cell surface markers can provide information on cell lineage, developmental stage, or function of particular cell populations.

T Lymphocytes (T Cells)

Macrophages process and present antigens that then activate T lymphocytes as part of cellular immunity mechanisms. The T cells may be subdivided in terms of function and the presence of the surface membrane antigens CD4 and CD8 into:

Helper cells (T_h) are CD4 positive or CD4+. These cells assist in B lymphocyte activation and synthesize interleukins 2 and 3, which in turn promote blood cell growth and maturation, and gamma interferon, which activates macrophages and GM-CSF.

Suppressor cells (T_s) counter the actions of T_h cells.

T_{tdh} lymphocytes secrete factors that activate and inhibit macrophage migration. These cells are involved in delayed hypersensitivity reactions.

Cytotoxic T cells (T_c) bind to target cells and release cytolytic substances.

The T cells constitute about 70% of circulating lymphocytes and proliferate and differentiate on contact with antigen presenting cells (such as macrophages and B cells). Some of the T cells become activated and mediate cellular immunity, reacting directly with antigens associated with cell membranes or by releasing lymphokines. Other T cells become activated after binding with antigens, and other T cells become helper cells. These helper T cells can proliferate and secrete lymphokines, which cause B cells to become plasma cells.

B Lymphocytes (B Cells)

These cells carry surface receptors that recognize and bind to foreign antigens, triggering cell proliferation and differentiation into plasma cells. The plasma cells synthesize and secrete antibodies against the inducing antigen. A small number of lymphocytes revert to a quiescent state and are called memory cells with surface immunoglobulin receptors because they are capable of mounting rapid responses in subsequent exposure.

OTHER CELLS

Monocytes, which become macrophages, are involved in both specific and innate immune reactions. **Mast cells** and **polymorphonuclear** (PMN) cells—neutrophils, eosinophils, and basophils—are involved in nonspecific defense mechanisms. Neutrophil PMNs have phagocytic activity, the **eosinophils** have cytotoxic functions, and the **basophils**, which become mast cells, release histamine and other chemicals/substances that initiate immediate hypersensitivity or local reactions to foreign substances. The Langerhans cells of the skin act as APCs and are involved in contact allergies.

The **natural killer** (NK) **cells** (historically described as null lymphocytes) are mainly large granular lymphocytes, and they mediate cytotoxicity against certain tumor- and virus-infected cells without requiring the presence of major histocompatibility complex (MHC) molecules on the target cells.

SOLUBLE MEDIATORS

Cytokines

The immune system is regulated by a variety of cytokines produced by the immune system, and these polypeptide cytokines can affect a number of cell types. The cytokines are proteins of small molecular masses of between 8 and 30 kDa (Table 7.1). These cytokines are produced in a wide variety of hemopoietic and nonhemopoietic cell types that are widely distributed. Some of the cytokine effects are autocrine, i.e., affecting the cells producing the cytokines, and other effects are paracrine, i.e., affecting cells neighboring the cytokine-producing cells, and they bind to cell surface receptors. These cytokine interactions can accentuate biological effects that may be initiated by a small change of one cytokine. Cytokine effects are not confined to immune responses and hematopoiesis; stress and inflammation also cause changes to cytokines.

These soluble proteins and peptides act at nano- to picomolar concentrations as humoral regulators. The generic term *cytokines* includes interleukins, lymphokines, monokines, interferons, colony-stimulating factors, and chemokines, and although the nomenclature was originally used to indicate production by specific cells

TABLE 7.1
Some Members of the Cytokine Family and Their Target Cells

Cytokine	Target
Interleukin 1	Hematopoietic and nonhematopoeitic cells
Interleukin 2	T cells, B cells, NK cells, macrophages
Interleukin 3	Stem cells, mast cells, granulocytes, monocytes/macrophages, eosinophils, megakaryocytes
Interleukin 4	B cells, T cells, mast cells, hematopoietic progenitor cells, monocyte/macrophage, hymocyte
Interleukin 5	Eosinophils, B cells
Interleukin 6	B cells, fibroblasts, thymocytes, T cells, hematopoeitic progenitor cells, hepatocytes
Interleukin 7	Pre-B cells, thymocytes
Interleukin 8	Neutrophils
Interleukin 9	T cells
Interleukin 10	Macrophages, T cells, neutrophils
Interleukin 12	NK cells
GM-CSF	Stem cells, monocytes/macrophages, eosinophils, neutrophils, endothelial cells
M-CSF	Stem cells, monocytes/macrophages
G-CSF	Stem cells, neutrophils
Tumor necrosis factor (TNF)	Tumor cells, fibroblasts, neutrophils, endothelial cells, monocytes/macrophages, chondrocytes, hepatocytes

(e.g., leukocytes/interleukins, lymphocytes/lymphokines, monocytes/monokines), it is now recognized that the cytokines have much wider roles.

Immunoglobulins

Immunoglobulins are produced by plasma cells (derived from B cells) and have specific antibody activities determined by their differing protein structures. There are five major immunoglobulin (Ig) classes—IgG, IgA, IgM, IgD, and IgE—with molecular masses ranging from approximately 150 kD (IgG) to 900 kD (IgM) (Table 7.2). The half-lives for IgE and IgD are shorter than those for the other immunoglobulins, whose half-lives are 10 to 20 days, and the plasma concentrations of both IgD and IgE are much lower than those of the other immunoglobulins (Bankert and Mazzaferro, 1999).

Complement Proteins

The plasma complement proteins represent about 10% of the plasma total globulins, and there are more than 30 individual proteins forming the complement system, which "complements" several immune functions, including cell lysis, adherence of

TABLE 7.2
Some Characteristics of Major Immunoglobulin Classes

Class	Major Function
IgG	Major antibody for toxins, viruses, and bacteria
IgA	Early antibacterial and antiviral defense in mucosal secretions
IgM	Major antibody formed after exposure to most antigens
IgD	Antigen receptor present on B cell surfaces
IgE	Present on mast cells and involved with immediate type of hypersensitivity

antibody-coated bacteria to macrophages, and immune modulation. The proteins interact with antigen–antibody complexes and cell membranes The complement system can be broadly divided into two pathways, the **classical pathway** and the **alternative pathway**, and the key converging focus protein is C3 (see Figure 7.2) (Holers et al., 1992; Burrell et al., 1992; Quimby, 1999). These two pathways are present in most common laboratory species, and the system is complex, but the complexity is added to by presence of several feedback mechanisms and the protein nomenclature. The complement proteins are numbered from 1 to 9 in their reaction sequence except for C4, and five proteins are designated by the letters B, D, P, H, and I.

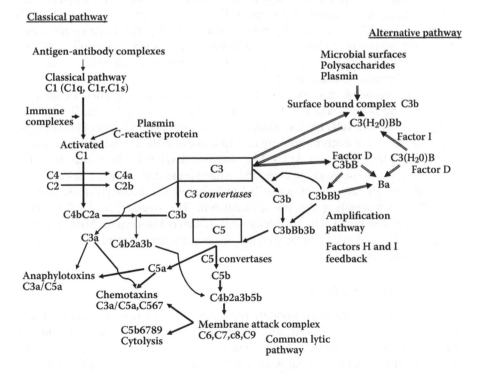

FIGURE 7.2 Classical (left) and alternative complement (right) pathways, with the common lytic and amplification pathways in the lower part of the figure.

IMMUNE RESPONSES

Some components of the immune system are involved with specific immune responses, whereas the responses of other components are nonspecific. The complex immune response system involves many interactions between cells and tissues that provide acquired and innate (natural) immunity (Table 7.3). Acquired or adaptive immune responses develop after the initial contact with pathogens; these responses may be divided into antibody-mediated responses and cell-mediated responses. The adaptive responses involve the T and B lymphocytes and their subsets. Following activation by antigen presenting cells (APCs), e.g., macrophages, the T cells release cytokines. The CD4+ cells can also be typed as **Th1** cells, which produce IL-2, interferon γ, and TNF-β as their main cytokines, and **Th2** cells, which produce the cytokines IL-4, IL-5, IL-6, IL-10, and IL-13. Both Th1 and Th2 cells produce IL-3, TNF-α, and GM-CSF. The released cytokines activate other T-cells and amplify cellular responses.

A large region of genes form the **major histocompatibility complex** (MHC), and these genes encode the proteins that control the antigen presentation and processing of the immune system. The MHC can be divided into three main classes, denoted by I, II, and III. Associated T and B cells engage with components of the molecules of MHC class II and **T cell receptors** (TCRs) to generate T-cell-dependent antibody responses. Innate immunity is dependent on the natural barriers of the skin and mucosae, and the actions of antimicrobial agents such as lysozyme, acute phase protein, and cytokines to foreign bodies before phagocytosis by neutrophils and monocytes. The processes of **phagocytosis** can be separated into several major phases: the migration of phagocytes to inflammatory sites via chemotaxis, followed by the attachment and ingestion of particles by the phagocytes, then the intracellular killing mechanisms (Benacerraf et al., 1959).

Immune response specificity is dependent on the amplification of antigen-selected T and B cells, and results from the interaction of the T cells, B cells, and **antigen presenting cells** (APCs). Dendritic cells are found in lymphoid tissues and interstitial tissues of nonlymphoid organs that express MHC II, and when these cells are activated they turn into APCs. Macrophages develop in the tissues from the monocytes, and the secretion patterns and antigen presentations of the macrophages vary with the different sites of origin. When activated, the monocytes-macrophages can kill neoplastic cells or alter tumor vasculature. The natural killer (NK) lymphocyte cells can lyse target cells via mechanisms that do not involve the MHC antigens; these NK cells are large granular lymphocytes that are not adherent and do not bear immunoglobulins.

HYPERSENSITIVITY AND ALLERGY

Four types of reactions are commonly recognized. Type I reactions occur rapidly after exposure (within minutes), with an antigen producing IgE antibodies on first exposure; subsequent exposure to the same antigen causes the release of histamine, serotonin, prostaglandins, heparin, etc., with varied clinical signs and symptoms.

TABLE 7.3

Simplified Division of Cells and Soluble Mediators of Innate and Acquired Immunity

Innate	Acquired/Adaptive
Polymorphonuclear neutrophils	T cells
Monocyte/macrophage	B cells
NK cells	Macrophages
	NK cells
Complement	Antibodies
Lysozyme	Cytokines
Acute phase proteins	
Interferon	
Cytokines	

Type II is characterized by cytolysis mediated through IgG or IgM antibodies, and targets the three main blood types causing anemia, leukopenia, and thrombocytopenia—this can also affect the progenitor cells. Type III, or Arthus, reactions are mediated mainly by IgG via the generation of antigen–antibody responses that can subsequently fix the complement. Type IV reactions are delayed hypersensitivity reactions mediated by T cells and macrophages, and with perivascular infiltration with monocytes, lymphocytes, and lymphoblasts.

AUTOIMMUNITY

The immune system can produce antibodies to endogenous antigens causing tissue injury. Autoimmunity is observed where the xenobiotic may act via the immune system by (1) phagocytosis of antibody-sensitized erythrocytes in some hemolytic anemias, or (2) direct effect on the cellular or humoral immune systems by covalent binding of the xenobiotic or metabolite to tissue macromolecules.

ASSAYS

Several immunological assays have been developed to evaluate the functional components of the immune system following exposure to xenobiotics. A small number of these assays will be mentioned here:

T and B cell enumeration
T-cell-dependent antibody response assay
NK cells
Oxidative burst and phagocytosis of monocytes/granulocytes

Immunoglobulins
Complement
Cytokines

In establishing and controlling these assays, compounds should be used as positive controls (both immunosuppressors and immunostimulants) to ensure the assays
are working satisfactorily and to enable comparisons between studies to be made.
The applications of monoclonal antibodies and flow cytometry are key technologies
for many of these assays.

FLOW CYTOMETRY AND ANTIBODIES

Flow cytometry is useful in assessing modulation of surface antigen expression,
respiratory burst cytokines, and enumeration of CD markers, particularly CD4:CD8
subsets (Burchiel et al., 1997, 1999; Gossett et al., 1999; Ormerod, 2000; Nguyen et
al., 2003).

ANTIBODIES

Although primarily the lymphocyte cluster of differentiation (CD) system addresses
human antibodies, there are increasing numbers of antibodies becoming available
for laboratory animals and veterinary animals (see appendix F). The allocated markers can be used to classify a variety of cell types, for example, T cells, B cells,
myeloid markers, macrophages, monocytes, etc. (Aasted et al., 1988; Jacobsen et
al., 1993; Cobbold et al., 1994; Cobbold and Metcalfe, 1994; Chabanne et al., 1994;
Galkowska et al., 1996; Neubert et al., 1996; Saint-Remy, 1997). Many of the suppliers of immunological reagents now provide suitable antisera or information on the
cross-reactivities of antisera with different species.

T AND B LYMPHOCYTES

These assays employ flow cytometry with the added potential of using four or five
different cell markers. Immunophenotyping of the lymphocytes is useful, but it is
not a functional assay. Absolute and percentage counts of cells can be obtained from
blood or splenic tissue, although reproducible sampling of splenic tissues can be
problematic. The development of antiviral drugs led to an enthusiastic approach to
the measurements of lymphocyte subsets, but these measurements in themselves are
not adequate for monitoring all of the potential effects on immune function; in addition to cell counts, functional tests are required.

Measurements of T and B lymphocyte subsets have been described for rodents,
dogs, and nonhuman primates (Chandler and Yang, 1981; Roholl et al., 1983; Eichberg et al., 1988; Evans et al., 1988; Jonker, 1990; Salemink et al., 1992; Tryphonas
et al., 1991; Bleavins et al., 1993; Ladics et al., 1993; Luster et al., 1993; Evans and
Fagg, 1994; Cardier et al., 1995; McKallip et al., 1995; Verdier et al., 1995; Jensen et
al., 1996; Foerster et al., 1997; Morris and Komocsar, 1997; Nam et al., 1998; Jones
et al., 2000; Yoshino et al., 2000; Lebrec et al., 2001; Reis et al., 2005).

The preanalytical variables that affect core leukocyte counts, e.g., sample collection procedures, diurnal variation, exercise, age stress, etc., also affect T and B lymphocyte measurements and other immune functions (Dhabhar et al., 1994; Pruett et al., 1993; Kingston and Hoffman-Goetz, 1996; Depres-Brummer et al., 1997) (see also Chapter 9).

T-Cell-Dependent Antibody Response (TDAR) Assays

These assays measure the effect of xenobiotics on the immune response to a cell immunogen, e.g., sheep red blood cells, keyhole limpet hemocyanin (KLH), or tetanus toxoid (TT), which can produce a robust antibody response.

Sheep Red Blood Cell (SRBC) Assay

Sheep red blood cells as a primary antigen are administered to rats and mice. The immune response is measured by either a plaque-forming cell assay or an enzyme-linked immunosorbent assay (ELISA) (van Loveren et al., 1991; Wood et al., 1992; Luster et al., 1993; Temple et al., 1993, 1995; Holsapple, 1995; Wilson et al., 1999). The plaque-forming cell (PFC) assay can measure the effect of a number of antibody-forming cells in the spleen. The ELISA method measures antibody levels in blood reflecting the IgM response from all of the lymphoid tissues. The preparation of the SRBC capture antigen is critical in this assay, and in establishing good comparative data between studies and laboratories.

Keyhole Limpet Hemocyanin (KLH) and Tetanus Toxoid (TT) Assays

These assays use alternative T-cell-dependent immunogens—either KLH or TT—because they are easier to standardize and prepare than SRBC, and they appear to be more reproducible (Exon and Talcott, 1995; Tryphonas et al., 2001; Gore et al., 2004).

TDAR tests are currently favored and may be applied to additional subgroups of animals but not to main treatment groups, as the additional procedures may confound the main purposes of the study. The immunoglobulin responses (IgG and IgM) are measured after immunization at appropriate time points. It would be advantageous if a single immunogen could be used for testing in both laboratory animals and humans, but this has yet to be achieved. KLH assays are not generally applied in early clinical trials due to potential allergic reactions observed in humans, and tetanus toxoid appears to be less reproducible in rodents. In these assays, there is a high interanimal variation, particularly in nonhuman primates, and serial measurements may be required to characterize the kinetics of the responses.

NK Cells

The activation of neutrophils can be demonstrated by monitoring CD11b positive cells using flow cytometry and whole blood, and this can be used as a marker of inflammation upregulation (Davis et al., 2000; Ruaux and Williams, 2000).

NK Cell Cytotoxic Function

The activity of NK cells (Munson and Phillips, 2000; Selgrade et al., 1992) can be measured by radiolabeling target lymphoma cells with chromium[51] isotope; the release of radioactivity from these cells when co-cultured with the effector splenic NK cells is then a measure of NK activity. Alternatively, a flow cytometric method can be used, as NK cells can be identified immunophenotypically as CD3+/CD16 or CD56+, with a small number being CD3+ CD56+ T cells. By mixing a source of target leukemic cells with isolated NK cells (effector cells) isolated from peripheral blood, reductions in the numbers of target cells labeled with DNA dye can be identified with the NK cell populations (identified by fluorescent CD labeling and flow cytometry). Ratios between effector and target cells can be determined (Cederbrant et al., 2003; Marcusson-Ståhl and Cederbrant, 2003). Natural killer cell activity assays and lymphocyte subsets phenotypic analysis may be acceptable alternatives to T-cell-dependent antigen assays in some cases.

Phagocytic Activity and Oxidative Burst Activity of Monocytes and Granulocytes

Blood is mixed with fluorescein-labeled opsonized bacteria, and the proportion of monocytes and granulocytes showing phagocytosis can be identified by flow cytometry.

For oxidative burst activity of monocytes and granulocytes, blood is mixed with opsonized bacteria in the presence of the activators (phorbol-12-myristate 13 acetate) a chemotactic peptide, and a fluorogenic rhodamine substrate. The degree of conversion of the substrate is related to the number of active phagocytic cells by flow cytometry. When reagent methods developed for human samples are applied to animal samples, there are some technical difficulties in rats where cells appear to be less active, and overall the data are variable.

Plasma Globulins

The plasma total globulins fraction is not usually measured directly, but it is derived from the plasma total protein and albumin values, and then expressed as an albumin: globulin (A:G) ratio. Major falls of total globulins generally indicate reductions of IgG, given the relative concentrations of this immunoglobulin class to the concentrations of other plasma immunoglobulins—IgA, IgM, and IgE.

Immunoglobulin Measurements

Methods include radial immunoassay, turbidometric, nephelometric, and ELISA immunoassays. For nonhuman primates, there are often close phylogenic similarities that allow reagents available for measurement of human proteins to be used, although there is an absence of homologous calibration materials. The availability of suitable reagents for other species is developing slowly, and for the immunoglobulins, IgG, IgA, and IgM antibodies are available for mouse, rat, and dog, but IgE methods are rarely available for laboratory animals (Day and Penhale, 1988; Caren and Brunmeier, 1987; Salauze et al., 1994; German et al., 1998; Bankert and Mazzaferro, 1999; Jones et al., 2000).

DIRECT IMMUNOGLOBULIN AND ANTIBODIES FOR ERYTHROCYTES, NEUTROPHILS, AND PLATELETS

Assays can be used to confirm immune-mediated mechanisms affecting these cell populations (see previous chapters), but it is not always possible to demonstrate the presence of the antibodies.

COMPLEMENT SYSTEM

Complement measurements have been made in several species, but measurements of individual complement proteins are rarely made in toxicology studies (Nelson et al., 1966; Holmberg et al., 1977; Horn et al., 1980; Ellingsworth et al., 1983). As a broad functional measurement of the complement, plasma samples are added to antibody-sensitized erythrocytes under standardized conditions, and the degree of erythrocytic lysis is measured. Immune complexes may also be measured in some species (Alexander et al., 1985; DeBoer and Madewell, 1983).

CYTOKINES

Methods and reagents are available for the measurement of tissue and plasma cytokines for several species, although reagents are not readily available for dog samples. Although cross-reactivity does occur where there is more than a 60% similar amino acid identity, the measurements of cytokines have some technical issues—not least is that plasma cytokines have relatively short half-lives of less than a few hours, and there are problems of standardization (Yamashita et al., 1994; Carson and Vignali, 1999; Hibino et al., 1999; Scheerlinck, 1999; Banks, 2000; Bienvenu et al., 2000; Pala et al., 2000; Foster, 2001; House, 2001; Pedersen et al., 2002; Kireta et al., 2005).

HOST RESISTANCE AND HYPERSENSITIVITY STUDIES

Studies for the prediction of reduced host resistance where rodents are challenged with bacterial, viral, or parasitic pathogens are not commonly used in toxicology. No single test is adequate, but predictivity is improved when combined with other tests. Host resistance and immunosuppression are both dependent on the amount of infectious agents used in the assays (Luster et al., 1993). In cases of hypersensitivity, there are well-established methods for evaluating contact sensitivity—the guinea pig maximization test, the Buehler test, and the murine local lymph node assay (Choquet-Kastylevski and Descotes, 1998; Hastings, 2001; CDER, 2002).

IMMUNOTOXICITY

A wide range of xenobiotics alter the immune system, which is similar in some respects to a target organ in that xenobiotic-induced injury to the immune system may follow a dose-response relationship with direct effects on the lymphoid organs or generalized immune dysfunctions. However, some immune-mediated effects cause toxicity indirectly, e.g., with mercury or gold, where the renal toxicity is secondary to the primary effects on the immune system. The presence of an antigen and the consequential immune response must be considered when interpreting immunotoxic effects. Xenobiotics may augment immunoreactivity by effects on specific immune cells, or indirectly by effects on immunoregulation, e.g., via the cytokines. There are also direct mechanisms involving interaction of a xenobiotic or its metabolites with a cell component of the immune system.

In evaluating immunotoxicity, the potential effects due to xenobiotics include unintentional **immunosuppression** resulting in resistance to the host defense mechanism, increased tumor incidence, or both; **immunogenicity**; **hypersensitivity**; and development of **autoimmunity.** Much less emphasis has been placed on adverse **immunostimulation**. Undesired effects on the immune system must be identified for drugs designed to suppress or stimulate the immune system in selected human or animal populations.

Immune-mediated destruction may affect all peripheral blood cell types or may be selective for one or more of the cell populations—erythrocytes, platelets, neutrophils. The changes of peripheral blood cell counts follow toxic changes of the bone marrow population, and the time of onset of adverse effects may be shorter when exposure is repeated in the same animal. If dosing stops, the peripheral blood should show evidence of recovery depending on the cell population life span, and the clearance or absence of the causative xenobiotic (or metabolite). Immune-mediated cytopenias are observed more frequently in dogs and nonhuman primates than in rodentia.

Compounds may be separated into two groups based on their molecular mass with respect to their potential to cause immunogenicity. Below a molecular mass of 10 kDa molecules may be immunogenic if they form hapten-protein complexes. Molecules with molecular mass of greater than 10 kDa, e.g., proteins, are often immunogenic in laboratory animals where the protein does not occur naturally or where similar proteins have markedly different structure. With some bioglogical therapeutic agents, animals will develop antibody responses that lead to neutralization and the presence of neutralizing antibodies. Several examples of drugs causing autoimmmune disorders have been mentioned in previous chapters, and these include three of the major cell types: hemolytic anemias with amoxicillin, neutropenia with methyl dopa, and pencillamine and thrombocytopenia with acetaminophen and quinine. Some compounds have toxic effects via cellular or humoral immunity, and in some cases these effects are mediated through covalent binding of a test compound or metabolite to tissue macromolecules.The development of antibodies to a xenobiotic can alter the pharmacodynamics and toxicity of a compound and the induced antibody responses, with antibodies being neutralizing, clearing, or reactive with endogenous proteins. These antibody responses are sometimes associated with drug hypersensitivity.

Unintentional immunological stimulation may cause adverse effects such as drug allergy, drug autoimmune reactions, and localized or systemic inflammatory reactions. These adverse effects are generally not predicted by standardized study designs, with the exception of drug-associated allergic contact dermatitis.

STRATEGIES FOR IMMUNOTOXICITY TESTING

As the science of immunotoxicology has progressed, there have been many proposed strategies for the detection of immunotoxic effects. With such a complex system, it is not surprising that some assays have been discarded because they were neither robust nor reproducible, or they failed to have sufficient predictive value. Some of the strategies and evaluations are referenced at the end of this chapter (Miller and Nicklin, 1987; Luster et al., 1988, 1992, 1994; Matsumoto et al., 1990; Sanders et al., 1991; Lebrec et al., 1994; Dean et al., 1994, 1998; Krzystyniak et al., 1995; Dean, 1997; Hastings, 1998; van Loveren et al., 1998; ICICIS, 1998; Pallardy et al., 1998; Karol and Stoliker, 1999; JPMA, 1999; EMEA, 2000a, 2000b, 2004; FDA, 2002; Putman et al., 2002). Included in these references are some recent regulatory and international recommendations (FDA, JPMA, and EMEA), but the discussions and wider application of tests for immunotoxic effects are by no means complete.

The development of biological therapeutic agents has brought additional challenges for immunotoxicological safety evaluations, e.g., proteins such as erythropoietin, monoclonal antibodies, and vaccines that activate immune systems. These therapeutic agents are potentially immunogenic, and although the focus may be on the effect of an agent on the target cell, and with knowledge gained from the testing of an agent against target cells *in vitro*, which may provide essential information on the toxic effects in cells, these tests may not anticipate the cascade of biological and toxic changes that may occur *in vivo* (Terrell and Green, 1994; Bussiere et al., 1995; Dayan, 1995; Thomas and Myers, 1995; Ryffel, 1997; Weirda et al., 2001).

When some extensively used immunotoxicity tests have been compared, problems have emerged concerning interlaboratory variation, standardization, and concordance between assays; sometimes the interlaboratory variations are greater than the expected biological effects with known test compounds. Running through many of these proposals, the common thread is that there is no single test suitable for all immunotoxic effects, and a battery of several tests is required. Several investigators have suggested a tiered approach relying on primary assessments, which are then followed by more specialized functional and immunological tests in separate and later studies.

The primary assessments should include the following, as they are commonly examined in toxicology studies:

Hematology—Full blood count including differential leukocyte counts
Plasma globulin and albumin/globulin ratios
Masses/weights—Body, spleen, thymus, kidney, liver
Histopathology—Spleen, thymus, lymph node cellularity, bone marrow
    cellularity

Plasma immunoglobulins may be used to provide additional information on globulin changes, although they are not particularly sensitive to mild immunosuppressions. Histopathology plays a major role in detecting effects on the immune system with examinations of the bone marrow, spleen, thymus, and lymph nodes. The examination of lymph nodes should include lymph nodes associated with the route of administration—gut-associated lymphoid tissue (GALT) for orally given compounds and bronchus-associated lymphoid tissue (BALT) and nasal-associated lymphoid tissue (NALT). The histopathological examinations are not without difficulties, in both common interpretation and assigning severity scores related to the findings (Kuper et al., 2000; Germolec et al., 2004).

As differences between species occur in both structure and function of the immune system (Haley, 2003), it is not surprising to find adverse effects in one species but no effects in another species. This makes both measurements and extrapolations of adverse immune effects problematic but essential for risk assessments (Selgrade et al., 1995; Buhles, 1998; Descotes, 2003; Snodin, 2004). Because biotechnology products have a reduced or no biological activity/immunogenic potential in some species, this can lead to poor predictivity for human exposures. Ideally, compounds should be tested in laboratory animals where the molecule has pharmacological activity.

Remembering that blood and immune systems are very dynamic, two main questions are: Which tests should be used? When? Although some evidence of immunosuppression can be obtained by regulatory standard/toxicological studies, other major organ toxicities and health status of animals should always be considered when there is a suspected effect on the immune system. It is important to exclude stress and potential pharmacological effects on the central nervous system as causes of apparent immunosuppression. Many of the efforts in developing immunotoxicity assays have been focused on cellular components rather than the soluble mediators, where changes are often transient in nature, such as the changes observed with cytokines.

Immune toxicity tests should always be considered where there is evidence of, or there is potential to cause, immunosuppression, or where compounds are highly concentrated in the immune systems. In immunological studies not all animals within a treatment group may be affected, and there may be no evidence of a dose-response relationship, although the frequency of adverse events is higher in the highest-dose treatment group. Immunological biomarkers ideally should be common to both humans and the laboratory animals used in preclinical testing, but this is not always possible. The inherent biological and analytical variations of immunotoxicity tests are much greater than the variations found for core hematology tests; these variations make the determinations of no-observed-effect level (NOEL) and no-observed-adverse-effect level (NOAEL) more difficult in risk assessment.

In summary, the following factors should be considered when examining xenobiotics for immunotoxic potential:

Does the compound have pharmacological properties that potentially could cause immunosuppression or immunostimulation, either by direct action on the immune system or via the central nervous system?
What are the cellular targets and mechanisms of action?
Is the compound or metabolite retained in the immune-related tissues at high concentrations?

Are there other toxicities or factors, e.g., stress, that are affecting the
immune system?
How many and which of the species dosed are affected?
What is the severity of the effects?
What is the dosage margin above the proposed therapeutic drug dose?
Are the effects reversible?
How do the results compare to known compounds that adversely affect the
immune system?

From this brief discussion of immunotoxicity, while histopathology has a dominant role, the reader should have come to recognize that hematology is also a key
contributor in identifying adverse events either by the simple identification of changes
in blood cells, which may be associated with bone marrow toxicity, or by erythrocytic immune hemolysis. There are expanding roles available to hematologists in
measuring the relevant cell populations, cellular functions, and soluble mediators,
such as complement proteins.

REFERENCES

GENERAL

Benacerraf, B., Sebestyen, M., and Copper, N. S. 1959. The clearances of antigen-antibody
complexes from the blood by the reticuloendothelial system. *J. Immunol.* 82:131–37.
Burrell, R., Flaherty, D. K., and Sauers, L. J. 1992. *Toxicology of the immune system: A
human approach.* New York: Van Nostrand Reinhold.
Descotes, J. 1999. *An introduction to immunotoxicology.* Philadelphia: Taylor & Francis.
Hadden, J. W., and Smith, D. L. 1992. Immunopharmacology: Immunomodulation and
immunotherapy. *J. Am. Med. Assoc.* 268:2964–69.
Kimber, I., and Dearman, R. J. 2002. Immune responses: Adverse versus nonadverse effects.
Toxicol. Pathol. 30:54–58.
Roitt, I. M., and Delves, P. J., eds. 1991. *Encyclopedia of immunology.* San Diego: Aca-
demic Press.

FLOW CYTOMETRY

Burchiel, S. W., Kerkvliet, N. L., Gerberick, F., Lawrence, D. A., and Ladics, G. S. 1997.
Assessment of immunotoxicity by multiparameter flow cytometry. *Fundam. Appl.
Toxicol.* 38:38–54.
Burchiel, S. W., Lauer, F. T., Gurule, D., Mounho, B. J., and Salas, V. M. 1999. Uses and
future applications of flow cytometry in immunotoxicity testing. *Methods* 19:28–35.
Gossett, K. A., Narayanan, P. K., Williams, D. M., Gore, E. R., Herzyk, D. J., Hart, T. K., and
Sellers, T. S. 1999. Flow cytometry in the preclinical development of biopharmaceuti-
cals. *Toxicol. Pathol.* 27:32–37.
Nguyen, D. T., Diamond, L. W., and Braylan, R. C. 2003. *Flow cytometry in hematopa-
thology: A visual approach to data analysis and interpretation.* Totowa, NJ: Humana
Press.
Ormerod, M. G., ed. 2000. *Flow cytometry: A practical approach.* 3rd ed., vol. 229. Oxford:
Oxford University Press.

ANTIBODIES

Aasted, B., Blixenkrone-Moller, M., Larsen, E. B., Bielfeldt-Ohmann, H., Simesen, R. B., and Uttenthal, A. 1988. Reactivity of eleven anti-human leucocyte monoclonal antibodies with lymphocytes from several domestic animals. *Vet. Immunol. Immunopathol.* 19:31–38.

Chabanne, L., Marchal, T., Kaplanski, C., Fournel, C., Magnol, J. P., Monier, J. C., and Rigal, D. 1994. Screening of 78 monoclonal antibodies directed against human leukocyte antigens for cross reactivity with surface markers on canine lymphocytes. *Tissue Antigens* 43:202–5.

Cobbold, S., Holmes, M., and Willett, B. 1994. The immunology of companion animals: Reagents and therapeutic strategies with potential veterinary and human clinical applications. *Immunol. Today* 15:347–53.

Cobbold, S., and Metcalfe, S. 1994b. Monoclonal antibodies that define canine homologues of human CD antigens: Summary of the First International Canine Leukocyte Antigen Workshop (CLAW). *Tissue Antigens* 43:137–54.

Galkowska, H., Waldemar, L. O., and Wojewodzka, U. 1996. Reactivity of antibodies directed against human antigens with surface markers on canine leukocytes. *Vet. Immunol. Immunopathol.* 53:329–34.

Jacobsen, C. N., Aasted, B., Broe, M. K., and Petersen, J. L. 1993. Reactivities of 20 anti-human monoclonal antibodies with leucocytes from ten different animal species. *Vet. Immunol. Immunopathol.* 39:461–66.

Neubert, R., Foester, M., Nogueira, A. C., and Helge, H. 1996. Cross-reactivity of antihuman monoclonal antibodies with cell surface receptors in the common marmoset. *Life. Sci.* 58:317–24.

Saint-Remy, J. M. 1997. Epitope mapping: A new method for biological evaluation and immunotoxicology. *Toxicology* 119:77–81.

T&B SUBSETS

Bleavins, M. R., Brott, D. A., Alvey, J. D., and de la Iglesia, F. A. 1993. Flow cytometric characterization of lymphocyte subpopulations in the Cynomolgus monkey (*Macaca fascicularis*). *Vet. Immunol. Immunopathol.* 37:1–13.

Cardier, J. E., Romano, E., and Soyano, A. 1995. Lipid peroxidation and changes in T lymphocyte subsets and lymphocyte proliferative response in experimental iron overload. *Immunopharm. Immuntoxicol.* 17:705–17.

Chandler, J. P., and Yang, T. J. 1981. Identification of canine lymphocyte populations by immunofluorescence surface marker analysis. *Int. Arch. Allergy Appl Immunol.* 65:62–68.

Eichberg, J. W., Montiel, M. M., Morale, B. A., King, D. E., Chanh, T. C., Kennedy, R. C., and Dreesman, G. R. 1988. Lymphocyte subsets in chimpanzees. *Lab. Anim. Sci.* 38:197–98.

Evans, G. O., and Fagg, R. 1994. Cellular and soluble CD4 measurements in Cynomolgus monkeys. *Exp. Anim.* 43:499–502.

Evans, G. O., Flynn, R. M., and Lupton, J. D. 1988. An immunogold labelling method for the enumeration of canine T-lymphocytes. *Vet. Quart.* 10:273–76.

Foerster, M., Delgado, I., Abraham, K., Gertsmayr, S., and Neubert, R. 1997. Comparative study of age dependent development of surface receptors on peripheral blood lymphocytes in children and young nonhuman primates (marmosets). *Life. Sci.* 60:773–85.

Jensen, J. E., Jensen, P. B., Nielsen, B., and Kemp, E. 1996. Flow cytometry in pan-T-, CD4+/-CD8+/-, and pan-B lymphocytes in blood samples obtained from healthy, nontreated rats: Comparison of lymphocyte subpopulations in blood samples obtained from rat heart and tail. *Lab. Anim. Sci.* 46:579–81.

Jones, R. D., Offutt, D. M., and Longmoor, B. A. 2000. Capture ELISA and flow cytometry method for toxicologic assessment following immunization and cyclophosphamide challenges in beagles. *Toxicol. Lett.* 115:33–44.

Jonker, M. 1990. The importance of non-human primates for preclinical testing of immunosuppressive monoclonal antibodies. *Semin. Immunol.* 2:427–36.

Ladics, G. S., White, K. L., Munson, A. E., Holsapple, M. P., and Morris, D. L. 1993. Separation of murine spenic B- and T-lymphocytes for use in immunological studies. *Toxicol. Methods* 3:143–56.

Lebrec, H., Kerdine, S., Gaspard, I., and Pallardy, M. 2001. Th_1/Th_2 responses to drugs. *Toxicology* 158:25–29.

Luster, M. I., et al. 1993. Risk assessment in immunotoxicology. II. Relationship between immune and host resistance tests. *Fundam. Appl. Toxicol.* 10:71–82.

McKallip, R. J., Nagarkati, M., and Nagarkatti, P. S. 1995. Immunotoxicity of AZT: Inhibitory effect on thymocyte differentiation and peripheral T cell responsiveness to gp120 of human deficiency virus. *Toxicol. Appl. Pharmacol.* 131:53–62.

Morris, D. L., and Komocsar, W. J. 1997. Immunophenotyping analysis of peripheral blood, splenic and thymic lymphocytes in male and female rats. *J. Pharmacol. Toxicol. Methods* 37:37–46.

Nam, K. H., Akari, H., Terao, L. K., Itagaki, S., and Yoshikawa, Y. 1998. Age related changes in major lymphocyte subsets in Cynomolgus monkeys. *Exp. Anim.* 47:159–66.

Reis, A. B., Carniero, C. M., das Graças Carvalho, M., Teixero-Carvalho, A., Giunchetti, R. C., Mayrink, W., Genaro, O., Corrêa-Oliviera, R., and Martins-Filho, O. A. 2005. Establishment of a microplate assay for flow cytometric assessment and its use for the evaluation of age-related phenotypic changes in canine blood whole blood leukocytes. *Vet. Immunol. Immunopathol.* 103:173–85.

Roholl, P. J. M., Leene, W., and Visser, J. W. M. 1983. T-cell differentiation in the rabbit. I. Cytofluorometric analysis of a T-cell antigen on lymphocytes from normal and dexamethasone-treated animals. *Thymus* 5:153–65.

Salemink, P., Doorstam, D., Maris, J., de Leeuw, P., van Doonmalen, A., and van de Berg, H. 1992. Analytical aspects of FACS (fluorescence activated cell sorter) identification of rat T and B peripheral lymphocytes in toxicity studies. *Comp. Haematol. Int.* 3:220–26.

Tryphonas, H., Luster, M. I., Schiffman, G., Dawson, L.-L., Hodgen, M., Germolec, D., Hayward, S., Bryce, F., Loo, J. C. K., Mandy, F., and Arnold, D. L. 1991. Effect of chronic exposure of PCB (Aroclor 1254) on specific and nonspecific parameters in the Rhesus (*Macaca mulatta*) monkey. *Fundam. Appl. Toxicol.* 16:773–86.

Verdier, F., Aujoclat, M., Condevaux, F., and Descotes, J. 1995. Determination of lymphocyte subsets and cytokine levels in Cynomolgus monkeys. *Toxicology* 105:81–90.

Yoshino, N., Ami, Y., Terao, K., Tashiro, F., and Honda, M. 2000. Upgrading of flow cytometric analysis for absolute counts, cytokines and other antigenic molecules of Cynomolgus monkeys (*Macaca fascicularis*) by using anti-human cross-reactive antibodies. *Exp. Anim.* 49:97–110.

STRESS/CIRCADIAN RHYTHM EFFECTS ON LYMPHOCYTES

Depres-Brummer, P., Bourin, P., Pages, N., Metzger, G., and Levi, F. 1997. Persistent T-lymphocyte rhythms despite suppressed circadian clock outputs in rats. *Am. J. Physiol.* 273: R1891–99.

Dhabhar, F. S., Miller, A. H., Stein, M., McEwen, B. S., and Spencer, R. L. 1994. Diurnal and acute stress changes in distribution of peripheral blood leukocyte subpopulations. *Brain Behav. Immun.* 8:66–79.

Kingston, S. G., and Hoffman-Goetz, L. 1996. Effect of environment enrichment and housing density on immune system reactivity to acute exercise stress. *Physiol. Behav.* 60:145–50.

Pruett, S. B., Ensley, D. K., and Crittenden, P. L. 1993. The role of chemical-induced stress responses in immunosuppression: A review of quantitative associations and cause-effect relationships between chemical-induced stress responses and immunosuppression. *J. Toxicol. Environ. Health* 39:163–92.

TDAR-SRBC

Holsapple, M. P. 1995. The plaque-forming cell (PFC) response in immunotoxicology: An approach to monitoring the primary effector function of B-lymphocytes. In *Methods in immunotoxicology*, ed. G. R. Burleston, J. H. Dean, and A. E. Munson, 71–108. Vol. 1. New York: Wiley-Liss.

Luster, M. I., et al. (13 authors) 1993. Risk assessment in immunotoxicology. II. Relationship between immune and host resistance tests. *Fundam. Appl. Toxicol.* 10:71–82.

Temple, L., Butterworth, L., Kawabata, T. T., Munson, A. E., and White, K. L. 1995. ELISA to measure SRBC-specific serum IgM: Method and data evaluation. In *Methods in immunotoxicology*, ed. G. R. Burleston, J. H. Dean, and A. E. Munson, Vol. 1. 137–59. New York: Wiley-Liss.

Temple, L., Kawabata, T. T., Munson, A. E., and White, K. L. 1993. Comparison of ELISA and plaque-forming cell assay for measuring the humoral immune response to SRBC in rats and mice treated with benzo[a]pyrene or cyclophosphamide. *Fundam. Appl. Toxicol.* 21:412–19.

van Loveren, H., Verlaaan, A. P. J., and Vos, J. G. 1991. An enzyme-linked immunosorbent assay of anti-sheep red blood cell antibodies of the classes M, G and A in the rat. *Int. J. Immunopharmacol.* 13:689–95.

Wilson, S. D., Munson, A. E., and Meade, J. 1999. Assessment of the functional integrity of the humoral immune response: The plaque forming cell assay and enzyme-linked immunosorbent assay. *Methods* 19:3–7.

Wood, S. C., Karras, J. G., and Holsapple, M. P. 1992. Integration of the human lymphocyte into immunotoxicological investigations. *Fundam. Appl. Toxicol.* 18:450–59.

TDAR-KLH/TT

Exon, J. H., and Talcott, P. A. 1995. Enzyme-linked immunosorbent assay (ELISA) for detection of specific IgG antibody in rats. In *Methods in immunotoxicology*, ed. G. R. Burleson, J. H. Dean, and A. E. Munson. Vol. 1. 109–24. New York: Wiley-Liss.

Gore, E. R., Gower, J., Kurali, E., Sui, J.-L., Bynum, J., Ennulat, D., and Herzyk, D. J. 2004. Primary antibody response to keyhole limpet hemocyanin in rats as model for immuntoxicity evaluation. *Toxicology* 197:23–35.

Tryphonas, H., Arnold, D. L., Bryce, F., Huang, J., Hodgen, M., Ladouceur, D. T., Fernie, S., Lepage-Parenteau, M., and Hayward, S. 2001. Effects of toxaphene on the immune system of Cynomolgus (*Macaca fasicularis*) monkeys. *Food Chem. Toxicol.* 39:947–58.

NK CELLS

Cederbrant, K., Marcusson-Ståhl, M., Condevaux, F., and Descotes, J. 2003. NK-cell activity in immunotoxicity drug evaluation. *Toxicology* 185:241–50.

Davis, H. M., Carpenter, D. C., Stahl, J. M., Zhang, W., Hynicka, W. P., and Griswold, D. E. 2000. Human granulocyte CD11b expression as a pharmacodynamic biomarker of inflammation. *J. Immunol. Methods.* 240:125–32.

Marcusson-Ståhl, M., and Cederbrant, K. 2003. A flow-cytometric NK-cytotoxicity assay adapted for use in rat repeated dose toxicity studies. *Toxicology* 193:269–79.

Munson, A. E., and Phillips, K. E. 2000. Natural killer cells and immunotoxicology. *Methods Mol. Biol.* 121:359–65.

Ruaux, C. G., and Williams, D. A. 2000. The effect of *ex vivo* refrigerated storage and cell preservation solution (Cyto-Chex II) on CD11b expression and oxidative burst activity of dog neutrophils, *Vet. Immunol. Immunopathol.* 74:56–69.

Selgrade, M. K., Daniels, M. J., and Dean, J. H. 1992. Correlation between chemical suppression of natural killer cell activity in mice and susceptibility to cytomegalovirus: Rationale for applying murine cytomegalovirus as a host resistance model and for interpreting immnotoxicty testing in terms of risk of disease. *J. Toxicol. Environ. Health* 37:123–37.

IMMUNOGLOBULINS

Bankert, R. B., and Mazzaferro, P. K. 1999. Biochemistry of immunoglobulins. In *The clinical chemistry of laboratory animals*, ed. W. F. Loeb and F. W. Quimby, 231–66. 2nd ed. Philadelphia: Taylor & Francis.

Caren, L. D., and Brunmeier, V. 1987. Immunotoxicology studies on octoxynol-9 and nonoxynol-9 in mice. *Toxicol. Lett.* 35:277–84.

Day, M. J., and Penhale, W. J. 1988. Serum immunoglobulin A concentrations in normal and diseased dogs. *Res. Vet. Sci.* 45:360–63.

German, A. J., Hall, E. J., and Day, M. J. 1998. Measurement of IgG, IgM and IgA concentrations in canine serum, saliva, tears and bile. *Vet. Immunol. Immunopathol.* 64:107–21.

Jones, R. D., Offutt, D. M., and Longmoor, B. A. 2000. Capture ELISA and flow cytometry methods for toxicologic assessment following immunization and cyclophosphamide challenges in beagles. *Toxicol. Lett.* 115:33–44.

Salauze, D., Serre, V., and Perrin, C. 1994. Quantification of total IgM and IgG levels in rat sera by a sandwich ELISA technique. *Comp. Haematol. Int.* 4:30–33.

COMPLEMENT

Alexander, N. J., Clarkson, T. B., and Fulgham, D. L. 1985. Circulating immune complexes in Cynomolgus macaques. *Lab. Anim. Sci.* 35:465–67.

Burrell, R., Flaherty, D. K., and Sauers, L. J. 1992. Effector mechanisms: Expressions of immunity. In *Toxicology of the immune system*, eds. R. Burrell, D. K. Flaherty, and L. J. Sauers, 79–98. New York: Van Nostrand Reinhold.

DeBoer, D. J., and Madewell, B. R. 1983. Detection of immune complexes in serum of dogs with neoplastic disease in solid-phase C1q binding using an enzyme-linked immunosorbent assay. *Am. J. Vet. Res.* 44:1710–13.

Ellingsworth, L. R., Holmberg, C. A., and Osburn, B. I. 1983. Hemolytic complement measurement in eleven species of non-human primate. *Vet. Immunol. Immunopathol.* 5:141–49.

Holers, V. M., Kinoshito, T., and Molina, H. 1992. The evolution of mouse and human complement C3-binding proteins: Divergence of form but conservation of function. *Immunol. Today* 13:231–36.

Holmberg, C. A., Ellingsworth, L., Osburn, B. I., and Grant, C. K. 1977. Measurement of hemolytic complement and the third component of complement in nonhuman primates. *Lab. Anim. Sci.* 27:993–98.

Horn, J. K., Ranson, J. H. C., Goldstein, I. M., Weisser, J., Curatola, D., Taylor, R., and Perez, H. D. 1980. Evidence of complement catabolism in experimental acute pancreatitis. *Am. J. Pathol.* 101:205–16.

Nelson, R. A., Jensen, J., Gigli, I., and Tamura, N. 1966. Methods for the separation, purification and measurement of nine components of hemolytic complement in guinea pig serum. *Immunochemistry* 3:111–35.

Quimby, F. W. 1999. Complement. In *The clinical chemistry of laboratory animals*, ed. W. F. Loeb and F. W. Quimby, 266–308. 2nd ed. Philadelphia: Taylor & Francis.

CYTOKINES

Banks, R. E. 2000. Measurement of cytokines in clinical samples using immunoassays: Problems and pitfalls. *Crit. Rev. Clin. Lab. Sci.* 37:131–82.

Bienvenu, J., Monneret, G., Fabien, N., and Revillard, J. P. 2000. The clinical usefulness of the measurement of cytokines. *Clin. Chem. Lab. Med.* 38:267–85.

Carson, R. T., and Vignali, D. A. A. 1999. Simultaneous quantification of 15 cytokines using a multiplexed flow cytometric assay. *J. Immunol. Methods* 227:41–52.

Foster, J. R. 2001. The functions of cytokines and their uses in toxicology. *Int. J. Exp. Pathol.* 82:171–92.

Hibino, H., et al. 1999. The common marmoset as a target preclinical primate model for cytokine and gene therapy studies. *Blood* 93:2839–48.

House, R. V. 2001. Cytokine measurement techniques for assessing hypersensitivity. *Toxicology* 158:51–58.

Kireta, S., Zola, H., Gilchrist, R. B., and Coates, P. T. H. 2005. Cross-reactivity of anti-human chemokine receptor and anti-TNF family antibodies with common marmoset (*Callithrix jacchus*) leukocytes. *Cell. Immunol.* 236:115–22.

Pala, P., Hussell, T., and Openshaw, P. J. 2000. Flow cytometric measurements of intracellular cytokines. *J. Immunol. Methods* 243:107–24.

Pedersen, L. G., Castelruiz, Y., Jacobsen, S., and Aasted, B. 2002. Identification of monoclonal antibodies that cross react with cytokines from different animal species. *Vet. Immunol. Immunopathol.* 88:111–22.

Scheerlinck, J.-P. Y. 1999. Functional and structural comparison of cytokines in different species. *Vet. Immunol. Immunpathol.* 72:39–44.

Yamashita, K., Fujinaga, T., Hagio, M., Izumisawa, Y., and Kotani, T. 1994. Canine acute phase response: Relationship between serum cytokine activity and acute phase protein in the dog. *J. Vet. Med. Sci.* 56:487–92.

HYPERSENSITIVITY

CDER. 2002. *Immunotoxicology evaluation of investigational new drugs*. Center for Drug Evaluation and Research, U.S. Department of Health and Human Services, Food and Drug Adminstration.

Choquet-Kastylevski, G., and Descotes, J. 1998. Value of animal models for predicting hypersensitivity reactions to medicinal products. *Toxicology* 129:27–35.

Hastings, K. L. 2001. Pre-clinical methods for detecting the hypersensitivity potential of pharmaceuticals: Regulatory considerations. *Toxicology* 158:85–89.

STRATEGIES FOR IMMUNOTOXICITY TESTING

Dean, J. H. 1997. Issues with introducing new immunotoxicology methods into safety assessment of pharmaceuticals. *Toxicology* 119:95–101.

Dean, J. H., Cornacoff, J. B., Haley, P. J., and Hincks, J. R. 1994. The integration of immunotoxicology in drug discovery and development: Investigative and *in vitro* possibilities. *Toxicol In Vitro* 8:939–44.

Dean, J. H., Hincks, J. R., and Remandet, B. 1998. Immunotoxicology assessment in the pharmaceutical industry. *Toxicol. Lett.* 102/103:247–55.

EMEA. 2000a. CPMP. *Note for guidance on non-clinical local tolerance testing.* Committee on Proprietary Medicinal Products.

EMEA. 2000b. *Note for guidance on repeated dose toxicity testing.* CPMP/SWP/1042/99. Committee on Proprietary Medicinal Products.

EMEA. 2004. *ICH S8 note for guidance on immunotoxicity studies for human pharmaceuticals.* London: European Medicines Agency.

FDA. 2002. *Guidance for industry. Immunotoxicology evaluation of investigational new drugs.* Center for Drug Evaluation and Research.

Hastings, K. L. 1998. What are the prospects for regulation in immunotoxicology? *Toxicol. Lett.* 102/103:267–70.

ICICIS. 1998. Report of validation study of assessment of direct immunotoxicity in the rat. The ICICIS, Group Investigators. International Collaborative Immunotoxicity Study. *Toxicology* 125:183–201.

JPMA. 1999. *International trends in immunotoxicity studies of medical products.* JPMA Drug Evaluation Committee Fundamental Research Group Data 92.

Karol, M. H., and Stoliker, D. 1999. Immunotoxicology: Past, present and future. *Inhalat. Toxicol.* 11:523–34.

Krzystyniak, K., Tryphonas, H., and Fournier, M. 1995. Approaches to chemical induced immunotoxicity. *Environ. Health Perspect.* 103(Suppl 9):17–22.

Lebrec, H., Blot, C., Pequet, S., Roger, R., Bohuon, C., and Pallardy, M. 1994. Immunotoxicological investigation using pharmaceutical drugs: *In vivo* evaluation of immune effects. *Fundam. Appl. Toxicol.* 23:159–68.

Luster, M. I., Munson, A. E., Thomas, P. T., Holsapple, M. P., Fenters, J. D., White, K. L., Lauer, L. D., Germolec, D. R., Rosenthal, G. J., and Dean, J. H. 1988. Development of a testing battery to assess chemical induced immunotoxicity. National Toxicology Program's guidelines for immunotoxicity evaluation in mice. *Fundam. Appl. Toxicol.* 10:2–19.

Luster, M. I., Pait, D. G., Portier, C., Rosenthal, G. J., Germolec, D. R., Comment, C. E., Munson, A. E., White, K., and Pollock, P. 1992. Qualitative and quantitative experimental models to aid in risk assessment of immunotoxicology. *Toxicol. Lett.* 65:71–78.

Luster, M. I., Portier, C., Pait, D. G., and Germolec, D. R. 1994. Use of animal studies in risk assessment for immunotoxicology. *Toxicology* 92:229–43.

Matsumoto, K., et al. 1990. Evaluation of immunotoxicity testing using azathioprine-treated rats: The International Collaborative Immunotoxicity Study (Azathioprine). *Eisei. Shikenjo. Hokoku* 108:34–39.

Miller, K., and Nicklin, S. 1987. Immunological aspects. In *The future of predictive safety evaluation*, ed. A. Worden, D. Parke, and J. Marks, 181–94. Vol. 1. Lancaster, England: MTP Press.

Pallardy, M., Kerdine, S., and Lebrec, H. 1998. Testing strategies in immunotoxicology. *Toxicol. Lett.* 102/103:257–60.

Putman, E., van Loveren, H., Bode, G., Dean, J., Hastings, K., Nakamura, K., Verdier, F., and van der Laan, J.-W. 2002. Assessment of the immunotoxic potential of human pharmaceuticals: A workshop report. *Drug Inf. J.* 36:417–27.

Sanders, V. M., Fuchs, B. A., Pruett, S. B., Kerkvliet, N. J., and Kaminski, N. E. 1991. Symposium on indirect mechanisms of immune modulation. *Fundam. Appl. Toxicol.* 17:641–50.

Van Loveren, H., De Jong, W. H., Vandebriel, R. J., Vos, J. G., and Garssen, J. 1998. Risk assessment and immunotoxicology. *Toxicol. Lett.* 102/103:261–65.

BIOTECHNOLOGY

Bussiere, J. L., McCormick, G. C., and Green, J. D. 1995. Preclinical safety assessment consideration in vaccine development. *Pharm. Biotechnol.* 6:61–79.

Dayan, A. D. 1995. Safety evaluation of biological and biotechnology-derived medicines. *Toxicology* 105:59–68.

Ryffel, B. 1997. Safety of human recombinant proteins. *Biomed. Environ. Sci.* 10:65–71.

Terrell, T. G., and Green, J. D. 1994. Issues with biotechnology products in toxicologic pathology. *Toxicol. Pathol.* 22:187–93.

Thomas, J. A., and Myers, L. A., eds. 1995. *Biotechnology and safety assessment.* New York: Raven Press Ltd.

Weirda, D., Smith, H. W., and Zwickl, C. M. 2001. Immunogenicity of biopharmaceuticals in laboratory animals. *Toxicology* 158:71–74.

PATHOLOGY AND SPECIES DIFFERENCES

Germolec, D. R., Nyska, A., Kashon, M., Kuper, C. F., Portier, C., Kommineni, C., Johnson, K. A., and Luster, M. I. 2004. Extended histopathology in immunotoxicity testing: Interlaboratory validation studies. *Toxicol. Sci.* 78:107–15.

Haley, P. J. 2003. Species differences in the structure and function of the immune system. *Toxicology* 188:49–71.

Kuper, C. F., Harleman, J. H., Richter-Reichelm, H. B., and Vos, J. G. 2000. Histopathologic approaches to detect changes indicative of immunotoxicity. *Toxicol. Pathol.* 28:454–66.

EXTRAPOLATION AND RISK ASSESSMENT

Buhles, W. C. 1998. Application of immunologic methods in clinical trials. *Toxicology* 129:73–89.

Descotes, J. 2003. From clinical to human toxicology: Linking animal research and risk assessment in man. *Toxicol. Lett.* 140/141:3–10.

Selgrade, M. J., Cooper, K. D., Devlin, R. B., van Loveren, H., Biagini, R. E., and Luster, M. I. 1995. Immunotoxicity—Bridging the gap between animal research and human health effects. *Fundam. Appl. Toxicol.* 24:13–21.

Snodin, D. J. 2004. Regulatory immunotoxicology: Does the published evidence support mandatory non-clinical immune function screening in drug development? *Reg. Toxicol. Pharmacol.* 40:336–55.

SIMPLE GLOSSARY

Allergen. An antigenic substance that can elicit an allergic response.

Anaphylaxis. An exaggerated reaction to an antigen after previous sensitization.

Anergy. Reduced reactivity to specific antigens.

Antigen. A substance that induces the formation of antibodies and reacts with a specific antibody; can be either external or internal to the body in origin.

Cytotoxicity. Producing a specific toxic effect on cells.

Sensitization. An allergic condition affecting skin or lungs on exposure to substance; further exposure even at lower levels may cause adverse reaction.

8 Hemostasis

Hemostasis is the term used to describe the normal physiological responses for the prevention and arrest of hemorrhage. The processes are dependent on interactions among the blood vessel walls, platelets, and coagulation factors that act to try to contain blood coagulation at the site of injury. Hemostasis is maintained by an equilibrium between coagulation factors (activators) and anticoagulation factors (inhibitors), and perturbations of this equilibrium will result in either hemorrhage (bleeding) or thrombosis (blood clots). The hemostatic mechanisms are basically similar for laboratory animals, but the mechanisms are complex. Several review articles are referenced for further reading at the end of this chapter.

Several of the regulatory documents referenced in Chapter 1 suggest the inclusion of "a measure of blood clotting time/potential" for toxicology studies. These documents do not specify the tests, so that many laboratories may elect to measure platelet counts and at least one measure of clotting potential—either prothrombin time or activated partial thromboplastin time—to meet these guidelines. In this chapter several additional tests are described that may be used in confirmatory or investigational studies.

Blood vessel walls have three concentric layers: the inner tunica intima layer with vascular endothelial cells containing Weibel–Palade bodies, the tunica media with variable elastic fibers and muscle cells, and the external adventitia, which contains collagen and muscle cells. These layers are separated by elastic lamina. Following injury to a blood vessel wall, the hemostatic mechanisms can be subdivided into vasoconstriction, platelet plug formation, and clot stabilization.

The endothelial cells of the intact blood vessel wall produce prostacyclin (which causes vasodilatation and inhibits platelet aggregation), antithrombin and thrombomodulin (an activator of protein C—both inhibit coagulation), and tissue plasminogen activator (t-PA; activates fibrinolysis). Following injury to the vessel wall, the membrane-bound tissue factor initiates coagulation and exposes collagen, which then allows platelets to bind to von Willebrand factor (vWF); this mediates platelet adhesion to the endothelium. Factor VIII is also transported by vWF in plasma (Figure 8.1).

Platelets are disc-shaped, nonnucleated blood cells produced from bone marrow megakaryocytes (see Chapter 2), and within the platelets there are alpha and dense granules that contain various constituents:

Alpha granules—platelet factor 4 (neutralizes heparin effects):
   Beta thrombomodulin
   Platelet-derived growth factor (chemotaxin for neutrophils, fibroblasts, and
      smooth muscle cells; mitogen for fibroblasts)
   von Willebrand Factor (adhesion molecules, complexes with factor VIII
      in plasma)

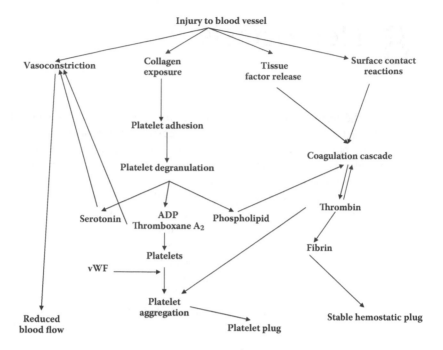

FIGURE 8.1 Mechanisms in vascular injury.

Thrombospondin (promotes platelet interaction)
Fibronectin (promotes adhesion of platelets and fibroblasts)

Dense granules—adenosine diphosphate (ADP; aggregation of platelets):
  Adenosine triphosphate (source of ADP)
  Serotonin (vasoconstriction)
  Calcium (coagulation and platelet function)

The platelets function by adhering to a surface, changing shape, releasing granular contents, and aggregating, and the adhesion process is via the platelet member receptors GpIb, GpIIb/IIIa, and vWF. The platelet shape changes from a discoid to irregular shapes with numerous cytoplasmic projections; in the early stages these shape changes are reversible, but at a later stage the shape changes do not reverse and the platelets become degranulated with the release of granular contents that increase platelet adhesion and aggregation.

Among the compounds released during platelet degranulation are ADP and thromboxane A_2. Adenosine diphosphate promotes platelet aggregation to form a primary hemostatic plug, and thromboxane A_2 produced by platelet prostaglandin synthesis potentiates platelet release and aggregation. Both serotonin and thromboxane A2 have vasoconstrictor effects.

The coagulation factors are mostly serine proteases, which are pro-enzymes and pro-cofactors, and these act in a sequential manner termed the coagulation cascade.

This cascade is often described as the intrinsic (or contact) and extrinsic (or tissue factor) pathways. These two pathways share common reactions, but the pathways are intertwined *in vivo* with feedback mechanisms and inhibitors.

The coagulation (clotting) factors are denoted by roman numerals:

Coagulation Factor	Common Synonym
I	Fibrinogen
II	Prothrombin
III	Tissue thromboplastin or tissue factor
IV	Calcium ions
V	Labile factor or proacclerin
VII	Stable factor or serum prothrombin conversion accelerator or proconvertin
VIII	Antihemophilic factor
IX	Christmas factor or plasma thromboplastin component
X	Stuart-Prower factor
XI	Plasma thromboplastin antecedent
XII	Hageman factor
XIII	Fibrin stabilizing factor or fibrinoligase

Other factors not denoted by roman numerals are prekallekrein (Fletcher factor) and high-molecular-weight kininogen (HMW kininogen; Fitzgerald factor). There is no factor VI. The coagulation factors are often denoted by *a* to indicate the activated form, e.g., XIIa. The life spans of these coagulation factors vary from a few hours (factor VII) to 8 to 10 days (fibrinogen). The coagulation pathways are shown in Figure 8.2 for endothelial surface reactions and Figure 8.3 for the further coagulation pathways.

In addition to feedback mechanisms, there are inhibitors for various steps of the coagulation cascade, and these include:

Activated proteins C and S, plasmin (also called fibrinolysin), which inhibit factors Va and VIIIa
Antithrombin III, which inhibits thrombin, IXa, Xa, and XIa
Heparin cofactor II
Tissue factor pathway inhibitor (TFPI), which inhibits the activity of the tissue factor VIIa complex and Xa
Alpha$_2$ macroglobulin, alpha$_2$ antiplasmin, and alpha$_2$ antitrypsin, which inhibit thrombin and circulating serine proteases

A majority of these proteins involved in coagulation pathways are synthesized by the liver. Factors II, VII, IX, and X and proteins C and S require vitamin K for their conversion from inactive to active factors and required functions.

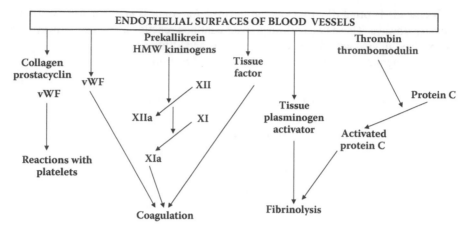

FIGURE 8.2   Endothelial surface mechanisms.

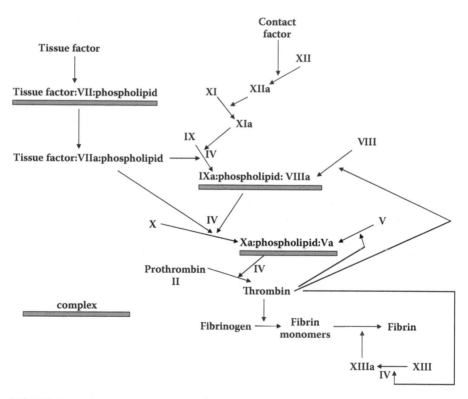

FIGURE 8.3   Coagulation factors and their cascade.

The lack of suitable materials means that most comparisons of factor activities between species are based on calibration standards that are designed for human plasma measurements, and not based on the species of laboratory animals. Fibrinogen (factor I) levels are similar for most species, being approximately 1.5 to 4 g/l. Compared to the activities of factors in humans:

Rat has higher activities of factors II, V, VIII, and XII, with lower activities of X and XI. There are also interstrain differences between Wistar and Sprague–Dawley rats (Lewis, 1996).

Mouse has higher activities of factors V, VII, and X, with lower activity for factor II.

Dog has higher activities of factors V, VII, VIII, IX, X, XI, and XIII, with lower activities for factor II.

Rabbit has higher activities of factors V, VII, VIII, IX, X, XI, and XIII, with lower activities for factor II.

Nonhuman primates have activities similar to those found in humans (Lewis, 1996; Dodds, 1997).

The fibrinolytic system degrades and dissolves circulating fibrin primarily by the action of plasmin, and the major components of this system are plasminogen, plasmin, activators, and inhibitors. Fibrinogen is an acute phase protein, and alterations of fibrinogen may reflect an acute phase reaction more often than coagulation dysfunction (Figure 8.4).

ANALYTICAL METHODS

In most toxicology studies, the assessment of coagulation generally is limited to four variables: platelet numbers, prothrombin and activated partial thromboplastin times, and fibrinogen. These four measurements are not sensitive enough to detect small changes in activation or inhibition of the coagulation systems, and they are essentially screening tests. Apart from the inclusion of platelet aggregation, the majority of measurements of other coagulation factors and platelet properties that are available for diagnostic purposes in human medicine remain largely absent from toxicology studies (Theus and Zbinden, 1984), although these additional tests are sometimes used to provide pharmacological information where compounds are designed to act on hemostatic mechanisms. The volumes of plasma and blood required for these additional tests remain a barrier to their utilization for small laboratory animals. Where the results of screening tests are similar to those of control group animals, but there is clear clinical evidence of a bleeding disorder, the use of other coagulation tests should be considered.

PLATELETS

These are counted by either fully automated counters as part of the full blood count, dedicated platelet analyzers, or manual counting of blood films. Automated platelet counters provide additional measurements, which include mean platelet volumes and

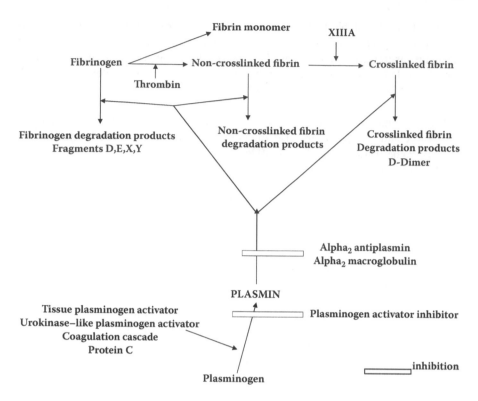

FIGURE 8.4   Fibrinogenolysis.

platelet distributions. Although other measurements (e.g, platelet crit) are provided by these counters, they require further evaluation for toxicology studies.

Platelet counts in dog, nonhuman primates, rabbit, and guinea pig are similar to those in humans (150 to 400 × 10⁹/l), but mouse and rat platelet counts are much higher (three- to fourfold higher than for humans). The approximate life spans of platelets are 4.5 days for mouse and rat and slightly longer (6 to 8 days) for rabbits and dogs. For comparison, the life span for human platelets is approximately 8 to 14 days. *In vivo* biotinylation techniques have been used to determine platelet life spans (Manning et al., 1996), and these *in vivo* techniques may have potential applications in the study of compounds affecting coagulation.

MEAN PLATELET VOLUME

Platelet volumes are usually expressed as a mean value of the platelet volume distribution (mean platelet volume [MPV]), and the volumes vary between species (Eason et al., 1986). Published canine MPV values appear to be more variable than most species, ranging from 6.7 to 13 fl; these differences are due in part to the greater variation of canine platelet shapes and the measuring technologies (Reardon et al., 1985, Eason et al., 1986; Evans and Smith, 1991). The platelet size is partly dependent on platelet age and thrombopoiesis, with a nonlinear relationship existing between platelet numbers and volumes. This relationship is disturbed under conditions of over- or underproduction of platelets.

COAGULATION CASCADE

Components of the coagulation cascade can be measured by a variety of turbidometric, nephelometric, chromogenic, enzyme-linked immunoassay, and immunological methods. The majority of these measurements are made using coagulometers/coagulyzers, or platelet aggregometers; reagents and analyzers may require adjustments to make them suitable for measurements of laboratory animal samples. Immunochemical assays developed for factor and enzyme inhibitor complexes may not be suitable for all species, as they are primarily designed for human medicine.

Some information on the factors affecting these assays is given in references and are provided at the end of this chapter, particularly for prothrombin and activated partial thromboplastin measurements. (Koepke et al., 1975; Mifsud, 1979; Hall, 1970; Zondag et al., 1985; Duncan et al., 1994; O'Brien et al., 1995; Tabata et al., 1995; Stringer and Seligmann, 1996; Adcock et al., 1997; Palm et al., Rob et al., 1997; Kurata et al., 1998); and Chapter 9 of this book. Preanalytical Variables and Chapter 10, Analytical Variables and Biosafety. Sampling procedures with small animals may lead to activation of coagulation factors due to the release of tissue factors. The citrate anticoagulant concentrations in the blood samples are critical for these assays.

PROTHROMBIN TIME (PT)

These measurements are largely based on two method principles: the Quick method (Quick et al., 1935) and the Owren method (Owren, 1959). The methods differ in that the Owren method requires a smaller sample volume than the Quick method, and the methods show differing sensitivities to the coagulation factors. The Owren method is more sensitive to factors II, VII, and X, while the Quick method is more sensitive for factors II, V, VII, and X and fibrinogen. The Quick method is also more sensitive to preanalytical variables such as the citrate anticoagulant concentration. The PT measurement is indicative of both the extrinsic and intrinsic pathway functionalities, and coagulation changes are more commonly observed as prolongations of the prothrombin time.

In human medicine the PT is measured and expressed as an international normalized ratio (INR); this expresses the measured prothombin time versus a mean value for a given population corrected by an international sensitivity index (ISI). The ISI value is assigned to a thromboplastin reagent by the manufacturer with reference to an international reference plasma. The formula for calculating INR is

INR = measured PT divided by mean PT of the normal range,
then multiplied by the *power* of the ISI value

Even several years after its introduction, most comparative studies or external assessment schemes show that absolute uniformity for INR between clinical laboratories using different analyzers and reagents has yet to be achieved (Pi et al., 1995). Currently there are no appropriate reference materials in animal hematology to allow INR determinations. (Expressing values for PT and APTT as a ratio of the mean normal or control values may occasionally be useful in detecting subtle changes and dose responses.)

Occasionally beagles may show factor VII deficiency with prolongations of prothrombin times (Mustard et al., 1962; Hovig et al., 1967). This finding is now largely of historical interest for toxicologists, although occasionally an affected individual may be found in a colony.

ACTIVATED PARTIAL THROMBOPLASTIN TIME (APTT)

The activated partial thromboplastin time primarily screens for factors IX, XI, XII, and VIII, prekallikrein, and high-molecular-weight kininogens and fibrinogen. Several activators are used in these measurements, including phopholipids of animal and plant origins, celite, silica, kaolin, and ellagic acid. Methods using kaolin and cephalin as activators are sometimes referred to as KCCT (kaolin/cephalin). The APTT measurement is indicative of the intrinsic pathway functionality.

BLEEDING TIME (BT)

A standardized and uniform cut on the skin or tail is made, and the time taken to stop bleeding is measured. The test must be performed in the animal care facility. It requires accurate timing and has an advantage of requiring very small blood volumes (Dejano et al., 1982).

WHOLE BLOOD CLOTTING TIME (WBCT)

Whole blood is added to glass tubes at 37 degrees Centigrade and the time taken for the blood to clot is measured. It offers a crude assessment of the intrinsic system but it is not widely used for toxicology studies.

ACTIVATED CLOTTING TIME (ACT)

Blood is mixed with a silaceous earth contact activator and the time taken for the blood to clot is measured. While it is suitable for the veterinary diagnostic clinic, it is not widely used for toxicology studies.

THROMBIN TIME (TT)

Exogenous thrombin is added to plasma and the time for clot formation is measured. This test assesses the terminal part of the common pathway. Prolongation of thrombin time indicates a deficiency of fibrinogen or inhibition of thrombin by fibrin degradation products (FDPs) or heparin.

FIBRINOGEN

This can be measured by the Clauss clotting rate assay where the clotting time is measured after the addition of thrombin and then referenced to known standards of plasma fibrinogen (Clauss, 1957; Gallimore et al., 1971; Lowe et al., 2004) or as fibrinogen concentrations derived from the clotting curve of the prothrombin time. The derived fibrinogen method has an advantage in that it does not require any additional plasma sample, but the results differ from those obtained using the Clauss assay and may be less accurate (Chitolie et al., 1994). High concentrations of fibrin degradation products in plasma may lead to an overestimation of fibrinogen when using clotting rate assay.

PLATELET FUNCTION TESTS

These tests include measurements of platelet aggregation with different agonists, platelet adhesion, and assessment of their granular contents. These tests should not be used as core tests, but they may be used where there is evidence of a perturbation of the coagulation pathway from either the core tests, clinical signs, or where the intended action of a compound is to alter coagulation. Usually, in the latter case, early compound development will include some *in vitro* assessment of platelet aggregation and function. Platelet function tests take careful standardization of the measuring conditions, and they are resource consuming due to sample preparation and analysis times. If appropriate, the potential of a compound for platelet aggregation may be assessed *in vitro* and *in vivo*.

Platelet functions are often assessed by **platelet aggregation** induced by added aggregating agents using either platelet-rich plasma or whole blood techniques (Born, 1962; Born and Cross, 1963; Cardinal and Flower, 1980). There is a continuing debate about the merits and differences observed using platelet-rich plasma versus whole blood samples, which require less sample preparation time, avoid the loss of larger platelets, and involve interactions due to leukocytes and erythrocytes present in whole blood (Sweeney et al., 1989; Hendra and Yudkin, 1990). In whole blood aggregometry the aggregation of platelets is detected by electrical impedance between detecting electrodes, while platelet aggregation is detected optically when using the platelet-rich plasma methods. A more recent approach is employed by the platelet function analyzer PFA 100™, which uses two cartridges—one cartridge with ADP and collagen and the second with collagen and epinephrine—through which blood is passed at high sheer force, and this detects abnormal platelet function (Keidel and Mischke, 1998; Kundu et al., 1995; Dyszkiewicz-Korpanty et al., 2005).

To induce platelet aggregation, five agents are commonly used: adenosine diphosphate (ADP), collagen, arachidonic acid, epinephrine, and ristocetin. The concentrations of the aggregating agents required for different species vary; for example, in the rat the concentration ranges of ADP required are lower, and collagen concentration ranges are higher than those used for humans. There are several other factors that affect analyses, including sex and hormonal influences (Emms and Lewis, 1984), and collagens from different sources appear to alter the responses to aggregation (Evans and Flynn, 1990).

There is a marked interindividual variation in the responses of guinea pigs to ADP (MacMillan and Sim, 1970), and several species show no apparent responses to some aggregating agents. These non-responses include epinephrine (adrenaline) with rat and guinea pig, and ristocetin with dog and rabbit (Sinakos and Caen, 1967; Hawkins, 1972; Feingold et al., 1986; Kurata et al., 1995).

OTHER FACTOR ASSAYS

Synthetic chromogenic peptide substrates now allow the measurements of several coagulation factors, inhibitors, and activators in human plasma. These factors include antithrombin III, protein C, protein S, tissue plasminogen activator, plasminogen activator, plasminogen activator inhibitor, $alpha_2$ antiplasmin, thrombomodulin, and fibrin monomers. There are several examples where chromogenic assays appear to work with samples from laboratory animals (e.g., $alpha_2$ antiplasmin, plasminogen, tissue plasminogen activator, and plasminogen activator inhibitor; Reilly et al., 1991; Schultze and Roth, 1995; Okazaki et al., 1998), but not all of the assays developed for human samples are suitable.

von Willebrand Factor

vWF (also termed factor VIII–related antigen) is an adhesive protein with a key role in hemostasis. This protein circulates in varied molecular forms ranging from small dimers to high-molecular-weight multimeric (HMM) forms. There are several methods for determining vWF; these include vWF antigen (vWF:ag), factor VIII coagulant assay (FVIII:c), ristocetin cofactor (vWf:RCo), collagen-binding capacity (vWF:CB), and Laurell rocket immunoelectrophoresis (Favaloro et al., 1995).

D-Dimer and Fibrinogen/Fibrin Degradation Products

In fibrinogenolysis and fibrinolysis, D-dimer is produced by the action of plasmin in the degradation of fibrin (Figure 8.4); this measurement is gradually being used as a replacement for fibrinogen degradation product (FDP) assays, which are sensitive to some fragments of degraded fibrinogen (Stokol, 2000; Stokol et al., 2003). Monoclonal antibodies are used to measure D-dimer by immunoturbidmetric or enzyme-linked immunosorbent assay (ELISA) methods, and the antibodies do not cross-react with other degradation products or all species (Reilly et al., 1991).

Vitamin K

The liver requires vitamin K for the synthesis of coagulation factors II, VII, IX, and X and proteins C and S, by activation via carboxylation of the glutamyl moieties in factors II, VII, IX, and X and proteins C and S. Vitamin K is a fat-soluble vitamin that occurs in three forms: K_1, phyloquinnone; K_2, menaquinnone; and K_3, menadione. This vitamin is synthesized by bacterial microflora in the ileum and colon before absorption via the lymphatics, and this absorption requires bile salts and fats. Analytical methods using high-performance liquid chromatography and mass spectroscopy are available, but are rarely necessary in toxicology studies (Kurata et al., 2005).

Calcium Factor IV

Plasma calcium or ionized calcium may be measured by several automated chemical methods or ion-selective electrodes under carefully controlled conditions. Profound hypocalcemia may affect the coagulation pathways. (Note that in many pathway diagrams for coagulation cascades, the role of calcium is usually indicated as Ca^{++} and not as factor IV.)

Other Techniques

Other methods used to measure the activation of coagulation, formation of platelet aggregates, e.g., thromboelastography, and rotation modifications of these techniques remain largely untested in laboratory animals.

TOXIC EFFECTS ON PLATELETS AND COAGULATION

Compounds may be deliberately designed to alter and correct hemostatic mechanisms, e.g., as in anticoagulant therapy or a rodenticide, but some xenobiotics can unexpectedly inhibit or stimulate the hemostatic mechanisms. More drugs have been alleged to cause thrombocytopenias than aplastic anemias or agranulocytosis, and thrombocytopenias are perhaps the most common of the blood dyscrasias reported due to xenobiotics.

The effects of hemostasis may be broadly divided into effects on platelet numbers and function, and those effects on the coagulation cascade, but there are connections between these two divisions in many events due to toxicity (Schrör, 1991). The timings of measurements may be critical in relation to dosing; for example, following a single dose of adriamycin given to rats, no changes in coagulation measurements were observed until 2 weeks after dosing (Poggi et al., 1979), which is in contrast to a case of disseminated intravascular hemolysis where measurements appeared to be altered only near the time of maximum dosing with a novel anticoagulant (Ganney and Brown, 1991). In toxicity affecting megakaryocytes, the circulating platelet count is reduced in about 5 to 7 days given the platelet life span in most laboratory animals.

The circulating platelet cell population numbers may be increased (**thrombophilia/thrombocytosis**) or reduced (**thrombocytopenia**). Platelets found attached to other platelets in groups are described as aggregated or clumped. The terms *partial* or *small clumps* may be used to indicate that a minority of platelets are aggregated. Some of the causes for artifactual changes as in pseudothrombocytopenia result from sample processing; these effects are discussed in Chapter 9.

Thrombocytosis is usually transient, slight, and asymptomatic; it may occur following acute blood loss, stress, exercise, mobilization of splenic pools, or as a rebound effect to thrombocytopenia. Thrombocytosis may be a chronic reaction to conditions such as chronic blood loss, chronic inflammation, severe iron deficiency, hemolytic anemias, and some cancers. The term *thrombocythemia* implies an overproduction of platelets due to stimulation of megakaryocyte production, which appears to be separate from the other hemopoietic processes. This may be observed in some myeloproliferative disorders. Where the platelet count is increased, the bone marrow examinations may help in determining if there is a reactive or myeloproliferative thrombocytosis. The presence of larger platelets in the blood generally indicates that there is increased production of platelets. Pseudothrombocytosis occurs rarely, sometimes with cryoglobulinemias where protein precipitates are counted as platelets.

Thrombocytopenia occurs due to increased destruction or increased utilization, reductions of platelet production, and decreased platelet survival. Platelet numbers must be considerably reduced before there is clinical evidence of platelet dysfunction, and platelet numbers alter more slowly than other blood cell types that have longer life spans, thus thrombocytopenias maybe caused by:

1. **Increased platelet utilization**. This occurs due to blood loss or injury to the vascular system.
2. **Blood loss**, such as that which occurs in severe gastrointestinal irritancy or systemic bleeding, can lead to thrombocytopenia. Here other blood measurements, clinical observations, and histopathological evidence can help in the interpretation of reduced platelet counts. Accelerated consumption of platelets can result from widespread activation of the endothelial injury or the coagulation system, e.g., in disseminated intravascular hemolysis. Minor bleeding due to toxicity may occur at various sites giving clinical and histological evidence, but with no apparent effects on the hemostatic measurements.
3. **Disseminated intravascular coagulation** (DIC). Endothelial cell damage leads to generalized activation of the coagulation and fibrinolytic pathways with widespread deposition of fibrin in the circulation and the formation of microthrombi. In DIC these microvascular thrombii may affect renal function, lead to further tissue necrosis, and cause gastrointestinal bleeding, and bleeding from intravenous injection or venepuncture sites. In addition to the thrombocytopenia, there are increased consumption of coagulation factors (giving prolonged PT and APTT), reductions of fibrinogen and with fragmentation of red cells, and increased fibrin degradation products. There is often histological evidence of fibrin deposition in the microvasculature, particularly in kidneys and lungs. In some instances DIC may proceed extremely rapidly and coagulation measurements may appear not to be altered, so timing of these measurements may be critical (Ganney and Brown, 1991).
4. **Reduced platelet production**. Platelet production falls primarily from effects on the bone marrow. Many xenobiotics that affect the erythroid and myeloid progenitor cells also affect the megakaryocytes, with pancytopenic effects in the blood. Thrombocytopenias may be associated with neoplasias in longer-term studies. In some thrombocytopenias, the bone marrow examination may show increased numbers of megakaryocytes, indicating the responsiveness of the bone marrow.

   Some example compounds are several cytostatic agents, thiazide diuretics, chloramphenicol, sulfonamides, and phenytoin.
5. **Immune-mediated thrombocytopenia**. In immune-mediated thrombocytopenias, the platelet life span (survival time) is decreased. The most common immune mechanism occurs when a compound forms an antigen with

a plasma protein, with the subsequent antibody formation to the new antigen. The antigen–antibody complexes then are absorbed onto the platelet surfaces, and the platelet-immune complexes are removed by macrophages, leading to a reduced platelet life span. Flow cytometric techniques can be useful in studying both *in vivo* and *in vitro* effects of immune-mediated thrombocytopenias.

Example compounds are quinine (Christie et al., 1985), heparin, and gold salts (Stavem et al., 1968).

6. **Altered platelet function**. Although the platelet count may be normal, prolongation of platelet aggregation and bleeding times may be associated with some drugs, including aspirin, other nonsteroidal anti-inflammatory compounds (NSAIDs), and some cephalosporins. Platelet function also may be altered in uremia caused by other drugs. Some compounds delay platelet aggregation times, e.g., oral contraceptives, while other compounds shorten platelet aggregation times, e.g., ethanol, penicillin, and phenyl butazone. There are several mechanisms where compounds can alter platelet function (Schrör, 1991); these include:

Interference with eicosanoid formation: The conversion of arachidonic acid may be blocked by compounds such as aspirin and some NSAIDs.

Alterations of platelet membrane stability: Some tricyclic antidepressants.

Reduction of receptor-mediator platelet activation: Some beta-lactam antibiotics, calcium antagonists, and calcium channel blockers.

Increase of receptor-mediator platelet activation: Streptokinase, ciclosporin.

EFFECTS ON THE COAGULATION PATHWAYS

Prothrombin and Activated Partial Thromboplastin Times

PT and APTT prolongations indicate decreased activity of one or more of the pro-coagulant factors of the coagulation cascade, but these tests are relatively insensitive to the activation of these pro-coagulation factors or effects on fibrinolysis.

Hepatoxicity

The liver synthesizes many of the proteins involved in coagulation homeostasis, and it requires major hepatotoxic effects to cause perturbations of this homeostasis due to the large reserve capacity of the liver. Following severe liver injury, synthesis of coagulation proteins is reduced with diminished clearance of activated coagulation factors and fibrin degradation products. This can lead to defects of coagulation with reduced platelet counts, prolonged PT and APTT, and increased fibrinolysis. Hepatic enzyme-inducing drugs can alter the metabolism of some anticoagulants when administered, and cause variations between species (Pritchard et al., 1987).

Vitamin K

Unless deliberately designed to induce vitamin K deficiency, laboratory animal diets contain enough vitamins to meet the required daily requirements, and the liver storage of the vitamin also provides an adequate supply for several days. Deficiencies can be caused by alterations in gut flora with antimicrobial agents, gastrointestinal toxicity and decreased absorption, altered biliary flow with hepatotoxic compounds, and increased vitamin K metabolism with some hepatic enzyme inducers. Estrogens may also affect vitamin L–dependent clotting factors (Hart, 1987).

The most common examples are the anticoagulants with coumarin moieties, e.g., warfarin, which inhibit the gamma decarboxylation of coagulation factors and vitamin K–dependent blood clotting factors. Again, changes of PT and APTT may occur at different time points following dosing (Owen and Bowie, 1978; Proudfoot, 1996; Kerins et al., 2002).

Fibrinogen

Changes of fibrinogen due to coagulation disorders occur less often than the changes related to inflammatory conditions and hepatotoxicity, as fibrinogen is an acute phase reactant. Increased fibrinogen values should be evaluated along with PT and APTT measurements to determine the cause for increased values.

OTHER COAGULATION FACTOR MEASUREMENTS

As suitable methods become more widely available, these measurements will find increasing usage either as efficacy biomarkers or for elucidating the more subtle effects on the coagulation pathways.

REFERENCES

GENERAL

Boon, G. D. 1993. An overview of hemostasis. *Toxicol. Pathol.* 21:170–79.
Broze, G. J. 1995. Tissue factor pathway inhibitor and the revised theory of coagulation. *Annu. Rev. Med.* 46:103–12.
Catalfamo, J. L., and Dodds, W. J. 1988. Hereditary and acquired thrombopathias. *Vet. Clin. N. Am. Small Anim. Prac.* 18:185–193.
Didisheim, P., Hattori, K., and Lewis, J. H. 1959. Hematologic and coagulation studies in various animal species. *J. Lab. Clin. Med.* 56:866–75.
Dodds, W. J. 1997. Hemostasis. In *Clinical biochemistry of domestic animals*, ed. J. J. Kaneko, J. W. Harvey, and M. L. Bruss, 241–84. 5th ed. San Diego: Academic Press.
Gibbons, J., and Mahaut, M. 2004. *Platelets and megakaryocytes.* Totawa, NJ: Humana Press.
Green, R. A. 1981. Hemostasis and disorders of coagulation. *Vet. Clin. N. Am. Small Anim. Prac.* 11:289–19.
Greene, C. E., Tsang, V. C., Prestwood, A. L., and Meriwether, E. A. 1981. Coagulation studies of plasma from healthy domesticated animals and persons. *Am. J. Vet. Res.* 42:2170–77.
Hall, D. E. 1972. *Blood coagulation and its disorders in the dog.* London: Balliere Tindall.
Lewis, J. H. 1996. *Comparative hemostasis in vertebrates.* New York: Plenum Press.

Lewis, J. H., and Didisheim, P. 1957. Differential diagnosis and treatment in hemorrhagic disease. *Arch. Intern. Med.* 100:157–68.

Lewis, J. H., van Thiel, D. H., Hasiba, U., Spero, J. A., and Gavaler, J. 1985. Comparative hematology and coagulation studies in rodentia (rats). *Comp. Biochem. Physiol.* 82A:211–15.

Littlewood, J. D. 1986. A practical approach to bleeding disorders in the dog. *J. Small Anim. Prac.* 27:397–409.

Marques, M. B., and Fritsma, G. A. 2006. *Quick guide to coagulation testing.* Washington, DC: AACC Press.

Mousa, S. A. 2004. *Anticoagulant, antiplatelets and thrombolytics.* Washington, DC: AACC Press.

Parry, B. W. 1985. Evaluation of haemostatic disorders in dogs and cats. In *Veterinary annual.* Washington, DC: J. Wright & Sons Ltd.

Spurling, N. W. 1981. Comparative physiology of blood clotting. *Comp. Biochem. Physiol.* 68A:541–48.

Theus, R., and Zbinden, G. 1984. Toxicological assessment of the hemostatic system: Regulatory requirements and industry practice. *Reg. Toxicol. Pharmacol.* 4:74–94.

Thomson, J. M., ed. 1991. *Blood coagulation and haemostasis: A practical guide.* 4th ed. London: Churchill Livingstone.

Platelets

Eason, C. T., Pattison, A., Howells, D. D., Mitcheson, J., and Bonner, F. W. 1986. Platelet population profiles: Significance of species variations and drug induced changes. *J. Appl. Toxicol.* 6:437–41.

Evans, G. O., and Smith, D. E. C. 1991. Platelet measurements in healthy beagles. *Comp. Haemtol. Int.* 1:49–51.

Manning, K. L., Novinger, S., Sullivan, P. S., and McDonald, T. P. 1996. Successful determination of platelet lifespan in C3H mice by in vivo biotinylation. *Lab. Anim. Sci.* 46:545–48.

Reardon, D. M., Hutchinson, D., Preston, F. E., and Trowbridge, E. A. 1985. The routine measurement of platelet volume: A comparison of aperture impedance and flow cytometric systems. *Clin. Lab. Haematol.* 7:251–57.

Prothrombin and Activated Partial Thromboplastin Times

Additional references are found in Chapter 9.

Adcock, D. M., Kressin, D. C., and Marlar, R. A. 1997. Effect of 3.2% vs 3.8% sodium citrate concentration on routine coagulation testing. *Am. J. Clin. Pathol.* 107:105–10.

Duncan, E. M., Casey, C. R., Duncan, B. M., and Lloyd, J. V. 1994. Effect of concentration of trisodium citrate anticoagulant on the calculation of the international normalized ratio and the international sensitivity index of thromboplastin. *Thromb. Haemost.* 72:84–88.

Evans, G. O., and Flynn, R. M. 1992. Activated partial thromboplastin time measurements in citrated canine plasma. *J. Comp. Pathol.* 106:79–82.

Hall, D. E. 1970. Sensitivity of different thromboplastin reagents to factor VII deficiency in the blood of beagle dogs. *Lab. Anim.* 4:55–59.

Hovig, T., Rowsell, H. C., Dodd, W. J., Jørgensen, L., and Mustard, J. F. 1967. Experimental hemostasis in normal dogs and dogs with congenital disorders of blood coagulation. *Blood* 30:636–68.

Koepke, J. A., Rodgers, J. L., and Ollivier, M. J. 1975. Pre-instrumental variables in coagulation testing. *Am. J. Clin. Pathol.* 64:591–96.

Kurata, M., Noguchi, N., Kasuga, Y., Sugimoto, T., Tanaka, K., and Hasegawa, T. 1998. Prolongation of PT and APTT under excessive anticoagulant in plasma from rats and dogs. *J. Toxicol. Sci.* 23:149–53.

Mifsud, C. V. 1979. The sensitivity of various thromboplastins in different animal species. *Pharmacol. Ther.* 5:251–56.

O'Brien, S. R., Sellers, T. S., and Meyer, D. J. 1995. Artifactual prolongation of the activated partial thromboplastin time associated with hemoconcentration in dogs. *J. Vet. Int. Med.* 3:163–70.

Owren, P. A. 1959. Thrombotest. A new method for controlling anticoagulant therapy. *Lancet* ii:754–58.

Palm, M., Frankenberg, L., Johanssen, M., and Jalkesten, E. 1997. Evaluation of coagulation tests in mouse plasma. *Comp. Haematol. Int.* 7:243–46.

Pi, D. W., Raboud, J. M., Filby, C., and Carter, C. J. 1995. Effect of thromboplastin and coagulometer interaction on the precision of the international normalised ratio. *J. Clin. Pathol.* 48:13–17.

Quick, A. J., Stanley-Brown, M., and Bancroft, F. W. 1935. A study of the coagulation defect in hemophilia and in jaundice. *Am. J. Med. Sci.* 190:501–11.

Rob, J. A., Tollefsen, S., and Helgeland, L. 1997. A rapid and highly sensitive chromogenic microplate for quantification of rat and human prothrombin. *Anal. Biochem.* 245:222–25.

Stringer, S. K., and Seligmann, B. E. 1996. Effects of two injectable anesthetic agents on coagulation assays in the rat. *Lab. Anim. Sci.* 46:430–33.

Tabata, H., Makamura, S., and Matsuzawa, T. 1995. Some species differences in the false prolongation of prothrombin times and activated partial thromboplastin times in toxicology. *Comp. Haematol. Int.* 5:140–44.

Zondag, A. C., Kolb, A. M., and Bax, N. M. A. 1985. Normal values of coagulation in canine blood. *Haemostasis* 15:318–23.

Factor VII Deficiency in Beagle Dogs

Hovig, T., Rowsell, H. C., Dodds, W. J., Jørgensen, L., and Mustard, J. F. 1967. Experimental hemostasis in normal dogs and dogs with congenital disorders of blood coagulation. *Blood* 30:636–68.

Mustard, J. F., Secord, D., Hoeksema, T. D., Downie, H. G., and Rowsell, H. C. 1962. Canine factor VII deficiency. *Br. J. Haemtol.* 8:43–47.

Bleed Time

Dejano, E., Villa, S., and de Gaetano, G. 1982. Bleeding time in rats: A comparison of different experimental conditions. *Thromb. Haemost.* 48:108–11.

Fibrinogen and Fibrinolysis

Chitolie, A., Mackie, I. J., Grant, D., Hamilton, J. L., and Machin, S. M. 1994. Inaccuracy of the derived fibrinogen. *Blood Coagul. Fibrinolysis* 5:955–57.

Clauss, A. 1957. Geringnungsphysiologische schnellmethode zur bestimmung des fibrinogens. *Acta Haematol.* 17:237–40.

Gallimore, M. J., Tyler, H. M., and Shaw, J. T. B. 1971. The measurement of fibrinolysis in the rat. *Thromb. Diath. Haemorrhag.* 26:295–310.

Lowe, G. D. O., Rumley, A., and Mackie, I. J. 2004. Plasma fibrinogen. *Ann. Clin. Biochem.* 41:430–40.

Platelet Aggregation

Born, G. V. 1962. Aggregation of blood platelets by adenosine diphosphate and its reversal. *Nature* 194:927–29.

Born, G. V., and Cross, M. J. 1963. The aggregation of platelets. *J. Physiol.* 168:178–95.

Cardinal, D. C., and Flower, R. J. 1980. The electronic aggregometer: A novel device for assessing platelet behaviour in blood. *J. Pharmacol. Methods* 3:135–58.

Dyszkiewicz-Korpanty, A. M., Frenkel, E. P., and Sarode, R. 2005. Approach to assessment of platelet function: Comparison between optical-based platelet-rich plasma and impedance-based whole blood platelet aggregation methods. *Clin. Appl. Thrombosis/Hemostasis* 11:25–35.

Emms, H., and Lewis, G. P. 1984. The influence of sex hormones on an experimental model of thrombosis in the rat. *Br. J. Pharm.* 81:71P.

Evans, G. O., and Flynn, R. M. 1990. Further observations on in-vitro aggregation of rat platelets with different collagens. *Thromb. Res.* 57:301–3.

Feingold, H. M., Pivacek, L. E., Melaragno, A. J., and Valeri, C. R. 1986. Coagulation assays and platelet aggregation patterns in human, baboon and canine blood. *Am. J. Vet. Res.* 47:2197–99.

Hawkins, R. I. 1972. The importance of platelet function tests in toxicological screening using laboratory animals. *Lab. Anim.* 6:155–67.

Hendra, T. J., and Yudkin, J. S. 1990. Whole blood platelet aggregation based on cell counting procedures. *Platelets* 1:57–66.

Keidel, A., and Mischke, R. 1998. Untersuchungen zur klinischen anwendung des plättchenfunktionsanalysen gerätes PFA-100 beim hund. *Berl. Munch Tierärztl. Wschr.* 111:452–56.

Kundu, S. K., Heilmann, E. J., Sio, R., Garcia, C., Davidson, R. M., and Ostgaard, R. A. 1998. Description of an in vitro platelet function analyser-PFA-100™. *Semin. Thromb. Hemostasis* 21:Suppl. 2 106–12.

Kurata, M., Ishikuka, N., Matsuzawa, M., Haruta, K., and Takeda, K. 1995. A comparative study of whole-blood platelet aggregation in laboratory animals: Its species differences and comparison with turbidometric method. *Comp. Biochem. Physiol.* 112C:395–9.

Macmillan, D. C., and Sim, A. K. 1970. A comparative study of platelet aggregation in man and laboratory animals. *Thromb. Diath. Haemorrh.* 24:385–94.

Sinakos, Z., and Caen, J. P. 1967. Platelet aggregation in mammalians (human, rat, rabbit, guinea-pig, horse, dog): A comparative study. *Thromb. Diath. Haemorrh.* 17:99–111.

Sweeney, J. D., Labuzzetta, J. W., Michielson, C. E., and Fitzpatrick, J. E. 1989. Whole blood aggregometry using impedance and particle counter methods. *Am. J. Clin. Pathol.* 92:794–97.

Other Factor Assays

Okazaki, M., Morio, Y., Iwai, S.-I., Miyamoto, K.-I., Sakamoto, H., Imai, K., and Oguchi, K. 1998. Age-related changes in blood coagulation and fibrinolysis in mice fed a high cholesterol diet. *Exp. Anim.* 47:237–46.

Reilly, C. F., Fujita, T., Mayer, E. J., and Siegfried, M. E. 1991. Both circulating and clot-bound plasminogen activator-1 inhibit endogenous fibrinolyis in the rat. *Arteriol. Thromb.* 11:1276–86.

Schultze, A. E., and Roth, R. A. 1995. Fibrinolytic activity in blood and lungs of rats treated with monocrotaline pyrrole. *Toxicol. Appl. Pharmacol.* 121:129–37.

vWF

Favaloro, E. J., Facey, D., and Grispo, L. 1995. Laboratory assessment of von Willebrand factor. Use of different assays can influence the diagnosis of von Willebrand's disease, dependent on differing sensitivity to sample preparation and differential recognition of high molecular weight vWF forms. *Am. J. Clin. Pathol.* 104:264–71.

D-Dimer

Reilly, C. F., Fujita, T., Mayer, E. J., and Siegfried, M. E. 1991. Both circulating and clot-bound plasminogen activator inhibitor-1 inhibit endogenous fibrinolysis in the rat. *Arterioscler. Thromb.* 11:1276–86.

Stokol, T. 2003. Plasma D-dimer for the diagnosis of thromboembolic disorders in dogs. *Vet. Clin. Small Anim.* 33:1419–35.

Stokol, T., Brooks, M. B., Erb, H. N., and Maudlin, G. E. 2000. D-dimer concentrations in healthy dogs and dogs with disseminated intravascular coagulation. *Am. J. Vet. Res.* 61:393–98.

Toxicity

Christie, D. J., Mullen, P. C., and Aster, R. H. 1985. Fab-mediated binding of drug-dependent antibodies to platelets in quinidine- and quinine-induced thrombocytopenia. *J. Clin. Invest.* 75:310–14.

Ganney, B. A., and Brown, G. 1991. Disseminated intravascular coagulation in the rat: A case history for a toxicological study. *Comp. Haematol. Int.* 1:172–77.

Hart, J. E. 1987. Vitamin K dependent blood clotting factors in female rats treated with oestrogens. *Thromb. Haemost.* 57:273–77.

Kerins, G. M., Endepols, S., and Magnicoll, A. D. 2002. The interaction between the indirect anticoagulant cocumatetralyl and calciferol (vitamin D3) in warfarin resistant rats (*Rattus norvigicus*). *Comp. Clin. Pathol.* 11:59–64.

Kurata, M., Iidaka, T., Yamasaki, N., Sasayama, Y., and Hamada, Y. 2005. Battery of test for profiling abnormalities of vitamin K-dependent coagulation factors in drug-toxicity studies in rats. *Exp. Anim.* 54:189–92.

Owen, C. A., and Bowie, E. J. W. 1978. Rat coagulation factors V, VII, XI and XII: Vitamin K dependent. *Haemostasis* 7:189–201.

Poggi, A., Kornblihtt, L., Delaini, F., Colombo, T., Mussoni, L., Reyers, I., and Donati, M. B. 1979. Delayed hypercoagulability after a single dose of adriamycin in normal rats. *Thromb. Res.* 16:639–50.

Pritchard, D. J., Wright, M. G., Sulsh, S., and Butler, W. H. 1987. The assessment of chemically induced liver injury in rats. *J. Appl. Toxicol.* 7:229–36.

Proudfoot, A. 1996. *Pesticide poisoning notes for guidance of medical practitioners.* London: Her Majesty's Stationery Office.

Schrör, K. 1991. Toxic influences on platelet function. *Arch. Toxicol. Suppl.* 14:147–52.

Stavem, P., Stromme, J., and Bull, O. 1968. Immunological studies in a case of gold salt induced thrombocytopenia. *Scand. J. Haematol.* 5:271–77.

9 Preanalytical Variables

In designing studies and before the interpretation of data, it is important to consider some of the variables that can affect data but are unrelated to the compound being tested. In addition, there are some general considerations to be made when choosing the best or most suitable methodology for the selected species. When there are several variables within a study, these effects may be additive, synergistic, or antagonistic, and furthermore, in toxicological studies the preanalytical and analytical variables can be modified or exaggerated by the pharmacological or toxicological effects of the test compound (and its metabolites). Some variables within a study or between studies may be tightly controlled, e.g., species and strain, while other variables may be less well controlled.

The preanalytical and analytical variations are much greater when dealing with laboratory animals compared to the effects observed on human data, and large preanalytical and analytical variations can limit the diagnostic sensitivity and specificity of the measurements made. In this chapter some of the preanalytical factors are considered, and they are listed here:

Animals—Species, strain, age, gender, pregnancy
Blood collection procedures—Volume, frequency, anticoagulant, collection site, anesthesia
Stress, environment, transportation
Nutrition and fluid balance
Chronobiological rhythms

ANIMALS: SPECIES, STRAIN, AGE, AND GENDER

In the previous chapters, attention has been drawn to some of the differences in the morphology of the various blood cells of the laboratory animal species, mean red cell volumes, and neutrophil:lymphocyte ratios.

The references for these variables—species, gender, and ages—are listed in appendix A. Referenced compendiums of data often show diverse values for single laboratory species obtained by different investigators, blood collection procedures, and analytical techniques.

STRAIN DIFFERENCE

Some of the differences between strains found within a particular study are no greater than data sets for the same strain obtained from different laboratories. Hematological differences have been reported for different rat strains (Ringler and Dabich, 1979; Lovell et al., 1981; Hackbarth et al., 1983; Pettersen et al., 1996), and between mouse strains, when using standardized blood collection procedures in each of these studies (Russell et al., 1951; Frith et al., 1980; Kajioka et al., 2000). Although most laboratory animals are now purpose bred and genetically more uniform, interstrain and breeder differences do occur.

Transgenic laboratory animals are mainly mice, which are useful to the preclinical scientists/pharmacologists, but for these investigations they remain a challenge, often with a reduced life span and dying in neonatal, fetal, or embryonic periods. The variability in blood measurements is greater in transgenic animals than for healthy common mouse and rat strains, and is often accompanied by unusual cell morphology. Given the small quantities of blood available during the early development periods and the difficulties associated with blood collection, any dilution procedures used to provide adequate amounts for automated blood counters will increase the variability of hematological measurements in these animals. Automated cell counts should always be checked microscopically when studying transgenic animals for the first time.

Inherited disorders occur in several species producing abnormal hematological profiles, but animals with these disorders are generally not used in toxicology studies. Rarely, beagle dogs with inherited factor VII deficiency appear (giving prolonged prothrombin times; see chapter 8), but most breeding colonies now appear to be free of this inherited disorder.

AGE-RELATED CHANGES

After birth, the mean corpuscular volume and reticulocyte count decrease toward values found in mature animals, and during this neonatal period the total leukocyte count increases as the immune system develops (Kojima et al., 1999). In rats there is an age-dependent increase of neutrophils, and this is accompanied by a decrease of lymphocytes after about 30 weeks of age (Cotchin and Roe, 1967; Bailly and Duprat, 1990; Edwards and Fuller, 1992), and similar changes are seen in guinea pigs (Kitagaki et al., 2005).

GENDER

The published data for most of the common laboratory animals are not consistent, with some authors claiming higher erythrocytic indices for males, and some authors claiming the opposite; very few papers claim leukocytic differences occur between sexes. Occasionally it may be useful to combine data for both genders where data are limited, but this must not disguise significant differences that may exist due to differing toxicities and metabolism between the two sexes. For example, there are marked differences in rats where males have higher plasminogen values than females (Oyekan and Botting, 1991), and platelet aggregation in mice (Duarte et al., 1986).

Although **reproductive toxicology** studies do not currently require the inclusion of hematology, there is an increasing use of core hematology measurements. During **pregnancy**, most publications agree that hemoglobin, hematocrit, and red cell blood counts decrease due to plasma expansion; the observations for leukocytic changes are varied, with some investigators finding reduced leukocyte counts and increased platelet counts, while others report no changes (Bortolotti, et al., 1989; Papworth and Clubb, 1995; Palm, 1997; LaBorde et al., 1999; Kim et al., 2000, 2002; also see Appendix A).

BLOOD COLLECTION

VOLUME AND FREQUENCY

Technical analytical developments during the last three decades have reduced the blood volumes required for complete or full blood cell counts, but the blood sample volume collected must be sufficiently large to be a representative sample. Using very small volumes may be attractive, but it can yield variable results, and this can be confirmed by the variations seen when taking a small number of serial samples, where they are often partially affected by tissue fluid contamination. Although estimates vary for circulating blood volumes in the various species, these estimates are based on different measurement techniques. A broad rule of 60 to 70 ml per kg for all species can be used to calculate the approximate blood volume in the body, but not all of this volume is available for collection, as it is "trapped" in tissues, etc.

Guidance for the removal of blood, including routes and volumes, is provided by several publications (McGuill and Rowan, 1989; BVA/FRAME/RSPCA/UFAW, 1993; Evans, 1994; Diehl et al., 2001). Toxicokineticists should also be aware of the effects of blood removal on their measurements, and their possible impact on combined toxicology studies (Hulse et al., 1981; Tamura et al., 1994). For animals weighing more than 1 kg, repetitive blood sampling is rarely a problem, but for smaller animals the guidance of collecting less than 15% of total blood volume for a single sampling and less than 7.5% per week for repetitive sampling is a reasonable guideline (McGuill and Rowan, 1989). Toxicology laboratories have local procedures covering permitted blood volumes and frequency of collection.

ANTICOAGULANTS

Ethylenediamine tetracetic acid (EDTA, or sequestrene) is the anticoagulant commonly used for full blood counts, whether as the di- or tripotassium salt of EDTA, and these prevent coagulation by binding calcium ions. Despite long-term usage and international recommendations concerning the use of EDTA, even this anticoagulant may be less than ideal as cellular changes and measurements occur over time (Reardon et al. 1991; ICSH, 1993). When the tripotassium salt is used as a liquid, the dilution of blood effectively lowers hemoglobin, mean corpuscular volume, red cell counts, and leukocyte counts by approximately 1–2% (Koepke et al., 1989). If either of these potassium salts is used at the appropriate concentration, and analysis is performed within a few hours of blood collection, the changes, if any, are very small (Goossens et al., 1991).

Coagulation measurements generally require a sample to be collected into sodium citrate as the anticoagulant, so for most studies two samples are required to perform core tests, i.e., EDTA and citrated tubes, and this increases the total blood volume requirements. Citrate forms a complex with calcium and prevents coagulation prior to analysis, and citrate solutions are used at two concentrations: 3.2% (105 mmol/l) and 3.8% (129 mmol/l) (Chantarangkul et al., 1998). Blood samples are diluted 1 part citrate solution to 9 parts blood, and after centrifugation the resultant plasma volumes are dependent on the initial hematocrit values (Duncan et al., 1994; Adcock et al., 1997).

Errors can occur if the tubes are incorrectly over- or underfilled (O'Brien et al., 1995; Kurata et al., 1998), or if the sample tubes are not stored correctly. Tubes containing liquid citrate have a short shelf life and should be stored upright in a refrigerator until required to prevent loss of citrate solution due to leakage or evaporation. Plasma should be separated within 2 hours of collection and analyzed within a few hours (GEHT, 1998).

EDTA and citrate samples should be thoroughly mixed immediately after collection, but not too vigorously, as this can lead to red cell fragmentation and hemolysis. To avoid hemolysis during sample collection and transfer to the sample tube, the needle size should be chosen to suit the animal and collection site, and blood should be transferred to the sample tube without excessive force.

The local laboratory should always be consulted as to the appropriate anticoagulant and sample collection tubes to be used. Non-core tests may require the use of other anticoagulants.

SAMPLE STORAGE

Overnight storage under refrigeration at 4 degrees Centigrade may introduce several artifacts affecting erythrocyte and leukocyte morphology, and reduce leukocyte and platelet counts. Some analyzers can give values that are similar for several measurements after storage for 2 to 3 days (Davies and Fisher, 1991; Byrne et al., 1994), but most analyzers detect the changes of blood morphology that occur during this storage period, and these changes are readily visible when blood films are prepared after storage for more than 24 hours. Following overnight storage, spherocytes and echinocytes can be observed in blood from rat and marmosets. Mean platelet volume and mean erythrocyte cell volume also change during storage (Evans and Smith, 1986; Byrne et al., 1994).

EDTA whole blood samples for hematology must never be shipped or stored frozen for core hematology tests (Hayashi et al., 1995). For coagulation studies, if analyses are delayed, then plasma samples should be stored at −40 or −80 degrees Centigrade.

BLOOD COLLECTION PROCEDURES

Blood collection procedures can have a significant impact on study results, and a variety of collection sites are used, particularly for the smaller laboratory animals. Blood sampling procedures require skilled personnel, and it is not uncommon to see a higher incidence of poor-quality specimens where routine procedures have not been used or where the operator is not sufficiently proficient. Some variations can be traced to individuals performing the collection procedures in slightly different ways, and therefore producing different results (Fowler, 1982). There are some circumstances where it is extremely difficult to obtain samples, e.g., from animals where the blood circulation is severely affected or in *extremis*.

Below is a list of sites used for blood collection in different species:

Dog: Jugular, femoral, saphenous, or cephalic veins
Ferret: Jugular or cephalic vein, or abdominal vena cava
Gerbil: Tail (coccygeal) vein, orbital-venous sinus, tail cut, or from the heart
Guinea pig: From the heart, jugular, or ear veins
Hamster: From the heart or jugular vein
Marmoset: Femoral vein or coccygeal vein
Mouse: Tail cut, tail vein, or from the heart
Monkey: Femoral, cephalic, jugular, saphenous veins, or from the heart
Rabbit: Ear vein or artery, jugular vein, or from the heart
Rat: Tail vein, retro-orbital venous plexus (sinus), sublingual vein, jugular
    vein, abdominal vena cava, abdominal aorta, and heart

Where blood cannot be collected from the central veins of smaller laboratory animals, sites such as the retro-orbital plexus and tail are used. Some of these procedures can only be carried out at termination, or may require anesthesia to allow interim sampling.

Numerous methods and data have been published for the collection of blood from rats and mice (Riley, 1960; Upton and Morgan, 1975; Cardy and Warner, 1979; Smith et al., 1968; Fowler et al., 1980; Gärtner et al., 1980; Archer and Riley, 1981; Nachtman et al., 1985; Suber and Kodell, 1985; Conybeare et al., 1988; Evans and Smith, 1991; Dameron et al., 1992;Wills et al., 1993; Matsuzawa et al., 1994; Bernardi et al., 1996; Walter, 1999; Mahl et al., 2000; Nahas et al., 2000; Pecaut et al., 2000; Schnell et al., 2002).

Samples collected from the retro-orbital plexus or tail may be contaminated by tissue fluid, and this can lead to results that are more variable than those obtained from central veins. Leukocyte counts are generally higher from these two sites compared to other sites. Carbon dioxide or gas-containing mixtures with carbon dioxide used for blood collection produce higher leukocyte counts than other techniques. There has been some controversy about the use of the retro-orbital plexus or tail for rat coagulation studies (Edwards and Fuller, 1993; Salemink et al., 1994; Edwards, 1994; Stringer and Seligman, 1996). In this author's experience, the coagulation results from these two sites are highly variable, and this is probably due to the local release of tissue pro-coagulant factors during the collection procedures. Effects due to anesthesia in other animals have been observed; for example, ketamine reduces erythrocytes and leukocytes in Rhesus monkeys (Bennett et al., 1992), and isoflurane has a similar effect in ferrets (Marini et al., 1994).

The choice of sampling procedures depends on the investigators and the local rules for animal welfare. The use of differing procedures has a marked effect on the hematology values. Multiple and excessive blood collections may in themselves cause hematological changes, e.g., anemia. Repeated or excessive blood sampling from the femoral veins in marmosets may lead to edema and hematoma at the site of collection. Even when available sample volumes are limited, pooling of samples prior to analysis should be avoided. Using more than one collection procedure within a study also should be avoided if possible, e.g., in rodent studies, collecting interim samples by tail collection procedures and the final sample by cardiac sampling at necropsy will result in differences when comparing the interim and final time of day due in part to the two collection procedures. These procedural differences emphasize the need to cautiously interpret data when making comparison between studies, and to establish reference ranges using your chosen procedures. Thus, erroneous results can be caused by poor sampling procedures, prolonged venous stasis, inadequate mixing with appropriate amounts of anticoagulant, and the presence of small blood clots.

STRESS, ENVIRONMENT, AND TRANSPORTATION

Physiological leukocytosis is well recognized as a response to stress, with the principle change being neutrophilia (pseudoneutrophilia) caused by movements in the distribution of neutrophils. Neutrophils present in the marginal compartments are flushed into the freely circulating compartments of the intravascular pool. While some individual animals have a quiet disposition and show minimal stress responses, the behavior of animals can be markedly affected by blood sampling procedures. Several of these changes can be attributed to stress-induced alterations of corticosteroids (Riley, 1981; Rowan, 1990).

Changing caging density can alter rat lymphocyte counts (Clausing and Gottschalk, 1989), mice lymphocytes (Peng et al., 1989) and rabbit hematological values (Whary et al., 1993). Changing kennel conditions can increase leukocyte counts, but hemoglobin and other red cell measurements may either increase of decrease (Soave and Boyle, 1965; Bickhardt et al., 1983; Kuhn and Hardegg, 1988). Changes may occur following severe exercise in dogs, and exercise effects may account for some of the variations observed between different environments (Panduranga et al., 1964; Reece and Wahlstrom, 1970).

Leukocytic variations, mainly neutrophilia, have been observed following shipping for rabbits (Toth and January, 1990), and for rats (Bean-Knudsen and Wagner, 1987) and dogs (Kuhn et al., 1991). Decreases of total leukocyte, T, and B lymphocyte counts following movements of mice within the same animal house have been reported (Drodowicz et al., 1990). Hemoglobin and hematocrit were reported to increase in rats after cage movements (Gärtner et al., 1980).

Light and dark cycles affect coagulation measurements in rats (Scheving and Pauly, 1967) and dogs (Andersen and Schalm, 1970) and eosinophil counts in rats (Bickhardt et al., 1983). These effects are probably due to changes of diurnal patterns. Increasing the animal room temperature can reduce leukocyte and hematocrit values of rats (Yamauchi et al., 1981), but hypothermia also can decrease leukocyte and increase hematocrit values. In severe toxicity, body temperature may alter, and this will affect the interpretation of data. Natural hibernation patterns in some species are mostly disturbed by study and animal room procedures.

Except where blood samples may be essential, for example, in the health screening for viral status on receipt of animals to new accommodation, it is preferable to allow a period of acclimatization prior to blood sampling to avoid some of the variations caused by housing changes, transportation, and dietary changes. During most studies, the animal room conditions will remain relatively constant, but stress effects in individual animals may occur and vary during toxicity studies, and their potential effects on hematological data should be recognized.

NUTRITION AND FLUID BALANCE

There is evidence of hemoconcentration with increases of hemoglobin, hematocrit, and red blood cell counts when food intake is reduced for 18- to 24-hour periods in rats, while leukocyte counts fall, which may reflect changes in the distribution between leukocyte pools (Apostoulou et al., 1976; Kast and Nishikawa, 1981; Pickering and Pickering, 1984, Maejima and Nagase, 1991; Levin et al., 1993). In dogs after an overnight fast, there is evidence of slight hemoconcentration. As it is the local practice in some laboratories to fast both rodents and dogs prior to blood sample collections, these possible short-term effects have been reviewed (Matsuzawa and Sakazume, 1994).

Water deprivation effects on hematological measurements have been reported for rats, hamsters, and gerbils but not guinea pigs (Kutscher, 1968; Clausing and Gottschalk, 1989), and dogs (Evans et al., 1991). Changes of parameters following severe dehydration in several species have been reviewed by Boyd (1981). Although the hematocrit is a convenient marker of hemodilution or hemoconcentration, it cannot be used to reliably correct for plasma protein or drug concentrations, because the red cells are confined to the vascular space. A plasma dilution factor calculated using pre- and posttreatment hemoglobin and hematocrit values gives a better estimate of effects due to blood loss, osmolality, and blood volume changes. The calculation of the plasma dilution factor is $Hb_{final}/Hb_{initial} \times (100 - Hct_{initial})/(100 - Hct_{final})$ (Flordal, 1995).

Severe dietary restrictions for periods longer than 2 weeks have caused reduction of lymphocytes and increases of erythrocyte count, hemoglobin, and hematocrit in rats (Schwartz et al., 1973; Oishi et al., 1979; Ogawa et al., 1985; Levin et al., 1993). Dietary restriction resulted in reductions of hemoglobin and red cell counts in Cynomolgus monkeys (Yoshida et al., 1994) and mice (Rathika et al., 2000).

Most of the references provided here are for studies where there have been deliberate manipulations of fluid and food intakes, which then have caused effects on hematological values. In association with several toxicity effects, animals may alter their food and water intakes, and these consumption data should be examined when interpreting hematological data. This is particularly important when there is evidence of gastrointestinal or renal toxicity.

CHRONOBIOLOGICAL EFFECTS

Most physiological measurements show some degree of change with time, either within the day or over longer periods. Some of these changes may be small, while others may be more obvious (Moore-Ede and Sulzman, 1981). There is evidence of diurnal variation of total leukocyte counts in most laboratory species (dogs:

Andersen and Schlam, 1970; Lilliehöök, 1997; rats and mice: Brown and Dougherty, 1956; Clark and Korst, 1969), and these rhythms extend to coagulation parameters (Scheving and Pauly, 1967; Cohen et al., 1978; Moser et al., 1996). Longer-term
rhythms in hematological values have been found in rats (Berger, 1983), dogs
(Andersen and Schlam, 1970; Sothern et al., 1993), and squirrel monkeys (Yoshida
et al., 1994), but were reportedly absent in captive stump-tailed macaques (De Neef
et al., 1987).

To reduce variations associated with time, it is sensible to consider taking samples at the same time of day if numbers permit, or to take samples in a randomized
order rather from all control group animals followed by all animals within the treatment groups.

SUMMARY

Some of the preanalytical variables are controllable, e.g., species, strain, age range,
and gender, when designing a study or series of studies. Good practice in animal care
minimizes stress and creates good sample collection procedures to reduce variations. Where blood samples are to be taken from several treatment groups containing
large numbers of animals, the order of sampling should be randomized; sampling
should also be randomized among collecting teams where more than one team is
involved with blood collection. Where the number of animals within a study is small
e.g., dog studies, samples should be taken to measure intra-animal variations before
and during the study. There is a need for careful examination of procedures and
data where blood collection procedures are altered or where baseline data appears
to markedly change.

REFERENCES

SPECIES, STRAIN, AGE, AND GENDER

These references are included Appendix A.

STRAIN

Frith, C. H., Suber, R. L., and Umholtz, R. 1980. Hematologic and clinical chemistry findings
     in control BALB/c and C57BL/6 mice. *Lab. Anim. Sci.* 30:835–40.
Kajioka, E. H., Andres, M. L., Nelson, G. A., and Gridley, D. S. 2000. Immunologic variables
     in male and female C57BL/6 mice from two sources. *Comp. Med.* 50:288–91.
Russell, E. S., Neufeld, E. F., and Higgins, C. T. 1951. Comparison of normal blood picture of
     young adults from 18, inbred strains of mice. *Proc. Soc. Expt. Biol. Med.* 78:761–6.

AGE

Bailly, Y. and Duprat, P. 1990. Normal blood cell values. Haemopoeitic system–pathophysi-
     ology. In: *Monographs on pathology of laboratory animals.* Jones T. C., Ward J. M.,
     Mohr U., and Hunt R. D. eds. Berlin, Germany: Springer-Verlag pp. 27–33.
Cotchin, E. and Roe, F. J. C. 1967. *Pathology of laboratory rats and mice.* Oxford, England:
     Blackwell Scientific Publications.

Edwards, C. J., and Fuller, J. 1992. Notes on age related changes in differential leucocyte counts of the Charles River Outbred albino SD rat and CD1 mouse. *Comp. Haematol. Int.* 2:58–64.

Kitagaki, M., Yamaguchi, M., Nakamura, M., Sakurada, K., Suwa, T. and Sasa, H. 2005. Age-related changes in haematology and serum chemistry of Weiser-Maples guineapigs. *Cavia porcellus. Lab. Anim.* 39: 321–30.

Kojima, S., Haruta, J., Enomoto, A., Fujisawa, H., Harada, T. and Maita, K. 1999. Age related changes and hematological changes in normal F344 rats during the neonatal period. *Exp. Anim.* 48:153–9.

GENDER

Oyekan, A. O. and Botting, J. H. 1991. Relationship between gender difference in intravascular aggregation of platelets and the fibrinolytic pathway in the rat. *Arch. Int. Pharmacodyn.* 39:321–30.

Duarte, A. P T., Ramwell, P., and Myers, A. 1986. Sex differences in mouse platelet aggregation. *Thromb. Res.* 43:33–9.

PREGNANCY—RATS

Kim, J.-C., Yun, H.-I., Kim, K.-H., Suh, J. E., and Chung, M.-K. 2000. Haematological values during normal pregnancy in Sprague-Dawley rats. *Comp. Haematol. Int.* 10:74–79.

LaBorde, J. B., Wall, K. S., Bolon, B., Kumpe, T. S., Patton, R., Zheng, Q., Kodell, R., and Young, J. F. 1999. Haematology and serum chemistry parameters of the pregnant rat. *Lab. Anim.* 33:275–87.

Papworth, T. A., and Clubb, S. K. 1995. Clinical pathology in the female rat during the pre- and postnatal period. *Comp. Haematol. Int.* 5:13–24.

PREGNANCY—RABBITS

Bortolotti, A., Castelli, D. and Bontati, M. 1989. Hematology and serum chemistry values of adult pregnant and newborn New Zealand rabbits (*Oryctolagus cuniculus*). *Lab. Anim. Sci.* 39:437–39.

Kim, J.-C., Yun, H.-I., Cha, S.-W., Kim, K.-H., Koh, W. S., and Chung, M.-K. 2002. Haematological changes during normal pregnancy in New Zealand white rabbits: A longitudinal study. *Comp. Haematol. Int.* 11:98–108.

Palm, M. 1997. Clinical pathology values in pregnant and non-pregnant rabbits. *Scand. J. Lab. Anim. Sci.* 24:177–83.

BLOOD VOLUMES AND COLLECTION METHODS

BVA/FRAME/RSPCA/UFAW Joint Working Group on Refinement. 1993. Removal of blood from laboratory mammals and birds. *Lab. Anim.* 27:1–22.

Diehl, K.-H., Hull, R., Morton, D., Pfister, R., Rabemampianina, Y., Smith, D., Vidal, J.-M., and van de Vorstenbosch, C. 2001. A good practice guide to the administration of substances and removal of blood including routes and volumes. *J. Appl. Toxicol.* 21:15–23.

Evans, G. O. 1994. Removal of blood from laboratory mammals and birds. *Lab. Anim.* 28:178–79.

Hulse, M., Feldman, S., and Bruckner, J. V. 1981. Effect of blood sampling schedules on protein drug binding in the rat. *J. Pharmacol. Exp. Ther.* 218:416–20.

McGuill, M. W., and Rowan, A. N. 1989. Biological effects of blood loss; implications for sampling volumes and techniques. *ILAR News* 31:15–18.

Tamura, A., Sugimoto, K., Sato, T., and Fujii, T. 1994. The effects of haematocrit, plasma protein concentration and temperature of drug-containing blood in-vitro on the concentrations of the drug in the plasma. *J. Pharm. Pharmacol.* 42:577–80.

Anticoagulants

Additional references are provided in chapter 8.

Adcock, D. M., Kressin, D. C., and Marlar, R. A. 1997. Effect of 3.2% vs 3.8% sodium citrate concentration on routine coagulation testing. *Am. J. Clin. Pathol.* 107:105–10.

Byrne, R. F., Andrews, C. M., Libretto, S. E., and Mifsud, C. V. 1994. Canine and feline haematology analysis: Comparative performance of Technicon H*1 and AVL MS8 VET analysers. *Comp. Haematol. Int.* 4:212–17.

Chantarangkul, V., Tripodi, A., Clerici, M., Negri, B., and Mannucci, P. M. 1998. Assessment of the influence of citrate concentration on the international normalized ratio (INR) determined with twelve reagent-instrument combinations. *Thromb. Haemost.* 80:258–62.

Davies, D. T., and Fisher, G. V. 1991. The validation and application of the Technicon H*1 for the complete automated evaluation of laboratory animal haematology. *Comp. Haematol. Int.* 1:91–105.

Duncan, E. M., Casey, C. R., Duncan, B. M., and Lloyd, J. V. 1994. Effect of concentration of trisodium citrate anticoagulant on the calculation of the international normalized ratio and the international sensitivity index of thromboplastin. *Thromb. Haemost.* 72:84–88.

Evans, G. O., and Smith, D. E. C. 1986. Further observations concerning MPV measurements. *Am. J. Clin. Pathol.* 86:126–27.

GEHT. 1998. Recommendations d'Etudes sur l'Hemostase et la Thrombose. Les variables pre analytiques en hemostase. *Sang. Thrombose. Valsseaux.* 10(Suppl.):3–21.

Goossens, W., van Duppen, V., and Verwilghen, R. L. 1991. K_2- or K_3-EDTA the anticoagulant of choice in routine haematology? *Clin. Lab. Haematol.* 13:291–95.

Hayashi, Y., Matsuzawa, T., Unno, T., Morita, N., and Nomura, M. 1995. Effects on haematology parameters during cold storage and cold transport of rat and dog samples. *Comp. Haematol. Int.* 5:251–55.

ICSH. 1993. Recommendations of the International Council for Standardisation in Haematology for ethylenediaminetetracetic acid anticoagulation of blood for blood cell counting and sizing. *Am. J. Clin. Pathol.* 100:371–72.

Koepke, J. A., van Assendelft, O. W., Bull, B. S., and Richardson-Jones, A. 1989. Standardization of EDTA anticoagulation for blood counting procedures. *Labmedica* 5:15–17.

Kurata, M., Noguchi, N., Kasuga, Y., Sugimoto, T., Tanaka, K., and Hasegawa, T. 1998. Prolongation of PT and APTT under excessive anticoagulant in plasma from rats and dogs. *J. Toxicol. Sci.* 23:149–53.

O'Brien, S. R., Sellers, T. S., and Meyer, D. J. 1995. Artifactual prolongation of the activated partial thromboplastin time associated with hemoconcentration in dogs. *J. Vet. Int. Med.* 9:169–70.

Reardon, D. M., Warner, B., and Trowbridge, E. A. 1991. EDTA, the traditional anticoagulant of haematology: With increased automation is it time for a review? *Med. Lab. Sci.* 48:72–75.

Blood Collection Procedures for Rats and Mice

Archer, R. K., and Riley, J. 1981. Standardised method for bleeding rats. *Lab. Anim.* 14:25–28.

Bernardi, C., Moneta, D., Brughera, M., Di Salvo, M., Lamparelli, D., Mazué, G., and Iatropoulos, M. J. 1996. Hematology and clinical chemistry in rats: Comparison of different blood collection sites. *Comp. Haematol. Int.* 6:160–66.

Cardy, R. H., and Warner, J. W. 1979. Effects of sequential bleeding on body weight gain in rats. *Lab. Anim. Sci.* 29:179–81.

Conybeare, G., Leslie, G. B., Angles, K., Barrett, R. J., Luke, J. S. H., and Gask, D. R. 1988. An improved simple technique for the collection of blood samples from rats and mice. *Lab. Anim.* 22:177–82.

Dameron, G. W., Weingand, K. W., Duderstadt, J. M., Odioso, L. W., Dierckman, T. A., Schwecke, W., and Baran, K. 1992. Effects of bleeding site on clinical laboratory testing of rats: Orbital venous plexus versus posterior vena cava. *Lab. Anim. Sci.* 42:299–301.

Evans, G. O., and Smith, D. E. C. 1991. Leucocyte values in rats and mice following carbon dioxide euthanasia. *Comp. Haematol. Int.* 1:233–35.

Fowler, J. S. L., Brown, J. S., and Flower, E. W. 1980. Comparison between ether and carbon dioxide anaesthesia for removal of small blood samples from rats. *Lab. Anim.* 14:275–78.

Fowler, J. S. L. 1982. Animal clinical chemistry and haematology for the toxicologist. *Arch. Toxicol. Suppl.* 5:152–59.

Gärtner, K., Buttner, D., Dohler, R., Friedel, R., Lindena, J., and Trautschold, I. 1980. Stress response to handling and experimental procedures. *Lab. Anim.* 14:267–74.

Mahl, A., Heining, P., Ulrich, P., Jakubowski, J., Bobadilla, M., Zeller, W., Bergmann, R., Singer, T., and Meister, L. 2000. Comparison of clinical pathology parameters with two different blood sampling techniques in rats: Retrobulbar plexus versus sublingual vein. *Lab. Anim.* 34:351–61.

Matsuzawa, T., Tabata, H., Sakazume, M., Yoshida, S., and Nakamura, S. 1994. A comparison of the effect of bleeding site on haematological and plasma chemistry values of F344 rats: The inferior vena cava, abdominal aorta and orbital venous plexus. *Comp. Haematol. Int.* 4:207–11.

Nachtman, R. G., Dunn, C. D. R., Driscoll, T. B., and Leach, C. S. 1985. Methods for repetitive measurements of multiple hematological parameters in individual rats. *Lab. Anim. Sci.* 36:505–8.

Nahas, K., Provost, J.-P., Baneux, Ph., and Rabemampianina, Y. 2000. Effects of acute blood removal via the sublingual vein on haematological and clinical parameters in Sprague-Dawley rats. *Lab. Anim.* 34:362–71.

Pecaut, M. J., Smith, A. L., Jones, T. A., and Gridley, D. S. 2000. Modification of immunologic and hematologic variables by method of CO_2 euthanasia. *Comp. Med.* 50:595–602.

Riley, V. 1960. Adaptation of orbital bleeding technic to rapid serial blood studies. *Proc. Soc. Exp. Biol. Med.* 104:751–54.

Schnell, M. S., Hardy, C., Hawley, M., Propert, K. J., and Wilson, J. M. 2002. Effect of blood collection technique in mice on clinical pathology parameters. *Hum. Gene Ther.* 13:155–62.

Smith, C. N., Neptun, D. A., and Irons, R. D. 1968. Effect of sampling site and collection procedure method on variations in baseline clinical pathology measurements in Fischer-344 rats. *Fundam. Appl. Toxicol.* 7:658–53.

Suber, R. L., and Kodell, R. L. 1985. The effects of three phlebotomy techniques on hematological and clinical chemical evaluation in Sprague-Dawley rats. *Vet. Clin. Pathol.* 14:23–30.

Upton, P. K., and Morgan, D. J. 1975. The effect of sampling technique on some blood parameters in the rat. *Lab. Anim.* 9:85–91.

Walter, G. L. 1999. Effects of carbon dioxide inhalation on hematology, coagulation and serum clinical chemistry values in rats. *Toxicol. Pathol.* 27:217–25.

Wills, J. E., Rowlands, M. A., North, D. C., and Evans, G. O. 1993. Effects of serial cardiac puncture and blood collection procedures in the rabbit. *Anim. Tech.* 44:39–52.

BLOOD COLLECTION PROCEDURES IN RAT COAGULATION STUDIES

Edwards, C. J. 1994. The orbital venous plexus as a sampling site for coagulation testing: The continuing saga. *Comp. Haematol. Int.* 4:236–38.

Edwards, C. J., and Fuller, J. 1992. Notes on age related changes in differential leucocyte counts of the Charles River Outbred albino SD rat and CD1 mouse. *Comp. Haematol. Int.* 2:58–64.

Edwards, C. J., and Fuller, J. 1993. Effect of bleeding site on coagulation results in the Charles River outbred albino SD rat. *Comp. Haematol. Int.* 3:40–42.

Salemink, P. J. M., Korsten, J., and de Leeuw, P. 1994. Prothrombin times and activated partial thromboplastin times in toxicology: A comparison of different blood withdrawal sites for Wistar Rats. *Comp. Haematol. Int.* 4:173–76.

Stringer, S. K., and Seligmann, B. E. 1996. Effects of two injectable anesthetic agents on coagulation in the rat. *Lab. Anim. Sci.* 46:430–33.

BLOOD COLLECTION PROCEDURES IN OTHER SPECIES

Bennett, J. S., Gossett, K. A., McCarthy, M. P., and Simpson, E. D. 1992. Effects of ketamine hydrochloride on serum biochemical and hematologic variables in Rhesus monkeys (*Macaca mulatta*). *Vet. Clin. Pathol.* 21:15–18.

Marini, R. P., Jackson, L. R., Esteves, M. I., Andrutis, K. A., Goslant, C. M., and Fox, J. G. 1994. Effect of isoflurane on hematologic variables in ferrets. *Am. J. Vet. Res.* 55:1479–83.

STRESS, ENVIRONMENT, AND TRANSPORTATION

Andersen, A. C., and Schalm, O. W. 1970. Hematology. In *The beagle as an experimental dog*, ed. A. C. Andersen and L. S. Good, 261–81. Ames: Iowa State University Press.

Bean-Knudsen, D. E., and Wagner, J. E. 1987. Effect of shipping stress on clinicopathologic indicators in F344/N rats. *Am. J. Vet. Res.* 48:306–8.

Bickhardt, K., Büttner, D., Müschen, U., and Plonait, H. 1983. Influence of bleeding procedure and some environmental conditions on stress-dependent blood constituents of laboratory rats. *Lab. Anim.* 17:161–65.

Clausing, P., and Gottschalk, M. 1989. Effects of drinking water acidification, restriction of water supply and individual caging on parameters of toxicological studies in rats. *Z. Versuchstierkd.* 32:129–34.

Drodowicz, C. K., Bowman, T. A., Webb, M. L., and Lang, C. M. 1990. Effect of in-house transport on murine corticosterone concentration and blood lymphocyte populations. *Am. J. Vet. Res.* 51:1841–46.

Gärtner, K., Büttner, D., Döhler, K., Friedel, R., Lindena, J., and Trautschold, I. 1980. Stress response of rats to handling and experimental procedures. *Lab. Anim.* 14:267–74.

Kuhn, G., and Hardegg, W. 1988. Effects of indoor and outdoor maintenance upon food intake, body weight and different blood parameters. *Z. Versuchstierkd.* 31:205–14.

Kuhn, G., Lichtwald, K., Hardegg, W., and Abel, H. H. 1991. Reaktionen von corticoiden enzymaktivitaten und haematologischen parametern auf transportstress bei hunden. *J. Exp. Anim. Sci.* 34:99–104.

Panduranga, R. V., Joshi, M. R., and Mahendar, M. 1964. Variations in the blood picture taken in the normal and splenectomised dogs after strenuous exercise. *Ind. Vet. J.* 41:529–32.

Peng, X., Lang, C. M., Drozdowicz, C. K., and Ohlsson-Wilhelm, B. M. 1989. Effect of cage density on plasma corticosterone and peripheral lymphocyte populations of laboratory mice. *Lab. Anim.* 23:302–6.

Reece, W. O., and Wahlstrom, J. D. 1970. Effect of feeding and excitement on the PCV of dogs. *Lab. Anim. Care* 20:1114–17.

Riley, V. 1981. Psychoneuroendocrine influence on immunocompetence and neoplasia. *Science* 212:1100–9.

Rowan, A. N. 1990. Refinement of animal research technique and validity of research data. *Fundam. Appl. Toxicol.* 15:25–32.

Scheving, L. E., and Pauly, J. E. 1967. Daily rhythmic variations in blood coagulation times in rats. *Anat. Rec.* 157:657–66.

Soave, O. A., and Boyle, C. C. 1965. A comparison of the hemograms of conditioned and non-conditioned laboratory dogs. *Lab. Anim. Care* 15:359–62.

Toth, L. A., and January, B. 1990. Physiological stabilization of rabbits after shipping. *Lab. Anim. Sci.* 40:384–87.

Whary, M., Peper, R., Borkowski, G., Lawrence, W., and Ferguson, F. 1993. The effects of group housing on the research use of the laboratory rabbit. *Lab. Anim.* 27:330–41.

Yamauchi, C., Fugita, S., Obara, T., and Ueda, T. 1982. Effects of room temperature on reproduction, body and organ weights, food and water intake and hematology in rats. *Lab. Anim. Sci.* 31:251–58.

NUTRITION AND FLUID BALANCE

Apostoulou, A., Saidt, L., and Brown, W. R. 1976. Effect of overnight fasting of young rats on water consumption, body weight, blood sampling and blood composition. *Lab. Anim. Sci.* 26:959–60.

Boyd, J. W. 1981. The relationships between blood haemoglobin concentration, packed cell volume and plasma protein concentration in dehydration. *Br. Vet. J.* 137:166–72.

Clausing, P., and Gottschalk, M. 1989. Effects of drinking water acidification, restriction of water supply and individual caging on parameters of toxicological studies in rats. *Z. Versuchstierkd.* 32:129–34.

Evans, G. O., Flynn, R. M., and Smith, D. E. C. 1991. Changes in differential leucocyte counts in male beagle dogs during water deprivation tests. *Comp. Haematol. Int.* 1:167–71.

Flordal, P. A. 1995. The plasma dilution factor: Predicting how concentrations in plasma and serum are affected by blood volume variations and blood loss. *J. Lab. Clin. Med.* 126:353–57.

Kast, A., and Nishikawa, J. 1981. The effect of fasting on acute oral toxicity of drugs in rats and mice. *Lab. Anim.* 15:359–64.

Kutscher, C. 1988. Plasma volume change during water deprivation in gerbils, hamsters, guinea pigs and rats. *Comp. Biochem. Physiol.* 25:929–36.

Levin, S., Semler, D., and Ruben, Z. 1993. Effects of two week feed restriction on some toxicological parameters in Sprague-Dawley rats. *Toxicol. Pathol.* 21:1–14.

Maejima, K., and Nagase, S. 1991. Effect of starvation and refeeding on the circadian rhythm of hematological and clinico-biochemical values and water intake of rats. *Exp. Anim.* 40:389–93.

Matsuzawa, T., and Sakazume, M. 1994. Effects of fasting on haematology and clinical chemistry values in the rat and dog. *Comp. Haematol. Int.* 4:152–56.

Ogawa, Y., Matsumoto, K., Kamata, E., Ikeda, Y., and Kaneko, T. 1985. Effect of feed restriction on peripheral blood and bone marrow counts of Wistar rats. *Exp. Anim.* 34:407–16.

Oishi, S., Ishi, H., and Hiraga, K. 1979. The effect of food restriction for 4 weeks on common toxicity parameters in male rats. *Toxicol. Appl. Pharmacol.* 47:15–22.

Pickering, R. G., and Pickering, C. E. 1984. The effects of reduced dietary intake upon the body and organ weights, and some clinical chemistry and haematological variates of the young Wistar rat. *Toxicol. Lett.* 21:271–77.

Rathika, V., Archunan, G., and Raghuraman, V. 2000. Influence of nutritionally induced anaemia on the oestrous cycle of mice. *Comp. Haematol. Int.* 10:90–93.

Schwartz, E., Tornaben, J., and Boxhill, G. C. 1973. The effects of food restriction on hematology, clinical chemistry and pathology in the albino rat. *Toxicol. Appl. Pharmacol.* 25:515–24.

Yoshida, T., Ohtoh, K., Narita, H., Ohkubo, F., Cho, F., and Yoshikawa, Y. 1994. Feeding experiment on laboratory-bred male Cynomolgus monkeys. II. Hematological and serum biochemical studies. *Exp. Anim.* 43:199–207.

CHRONOBIOLOGICAL EFFECTS

Berger, J. 1983. The effect of repeated bleedings on bone marrow and blood morphology in adult laboratory rats. *Folia Haematol.* 110:685–91.

Brown, H. E., and Dougherty, T. F. 1956. Diurnal variation of blood leukocytes in normal and adrenalectomized mice. *Endocrinology* 56:365–75.

Clark, R. H., and Korst, D. R. 1969. Circadian periodicity of bone marrow, mitotic activity and reticulocyte counts in rats and mice. *Science* 166:236–37.

Cohen, M., Simmons, D. J., and Joist, J. H. 1978. Diurnal hemostatic changes in the rat. *Thromb. Res.* 12:965–71.

De Neef, K. J., Nieuwenhuijsen, K., Lammers, A. J. J. C., Degen, A. J. M., and Verbon, F. 1987. Blood variables in adult stumptail macaques (*Macaca artoides*) living in a captive group: Annual variability. *J. Med. Primatol.* 16:237–47.

Lilliehöök, I. 1997. Diurnal variation of canine blood leukocyte counts. *Vet. Clin. Pathol.* 26:113–17.

Moore-Ede, M. C., and Sulzman, F. M. 1983. Internal temporal order. In *Handbook of behavioural neurobiology, biological rhythms*, ed. J. Aschoff, 215–41. Vol. 4. New York: Plenum Press.

Moser, J., Meyers, K. M., Meinkoth, J. H., and Brassard, J. A. 1996. Temporal variation and factors affecting measurement of canine von Willebrand factor. *Am. J. Vet. Res.* 57:1288–93.

Scheving, L. E., and Pauly, J. E. 1967. Daily rhythmic variations in blood coagulation times in rats. *Anat. Rec.* 157:657–66.

Sothern, R. B., Farber, M. S., and Gruber, S. A. 1993. Circannual variations in baseline blood values of dogs. *Chronobiol. Int.* 10:364–82.

10 Analytical Variables and Biosafety

The variations due to analytical variables in hematological measurements usually are much less than variations due to preanalytical factors, i.e., the biological components and the sample collection procedures discussed in the previous chapter. In earlier chapters, the differences between blood cell numbers, shapes, and sizes for laboratory animals have been mentioned, and not all of the analytical equipment and reagents used in human medicine are appropriate for use with animal species.

Most hematology analyzers are effectively particle (or cell) counters with a chemical method for determining hemoglobin. A wide variety of hematology analyzers are available, including some that are capable of performing erythrocytic measurements, total differential leukocyte counts, reticulocyte counts, platelet counts, and separately coagulation factors. A few of the analyzer manufacturers offer dedicated software algorithms for counting blood cells in selected animal species, but others recognize that their analyzers are not appropriate for all laboratory animals or cannot provide full blood counts. Platelets and reticulocytes may be counted using dedicated stand-alone analyzers, or these counts may be combined within one analyzer. Currently there are a few computerized image analysis systems designed for cell recognition and the counting of human blood films, but these have not been validated for laboratory animals, although in the past at least one system was adapted for use with laboratory animal films. For most automated analyzers, there is a minimum requirement of approximately 80 to 150 µl of blood; for coagulation assays, approximately 50 to 150 µl of plasma is required, but requirements vary for each analyzer and laboratory.

Different cell counting technologies include (1) electrical impedance, where cells are counted by passage through an electrical current—hence impedance of the current, optical, polarized, or laser light scattering to determine cell size and complexity, or (2) cytochemical or fluorescence staining for leukocyte differential counts (Lombarts et al., 1986; Schoentag, 1988; Paterakis et al., 1994; Groner and Simson, 1995). Most counters use hydrodynamic focusing to stream the cells past the detection systems. Analyzers must be adjusted either manually or by on-board programs to select the thresholds for counting the differently sized cells. Although some analyzers produce a three-part differential leukocyte count, e.g., monocytes, granulocytes, and lymphocytes, these are generally not regarded as adequate for today's requirements for a five-part differential leukocyte count, i.e., lymphocytes, neutrophils, eosinophils, basophils, and monocytes. The counting and sizing of animal cells using automated analyzers is made difficult by the varied cell sizes, shapes, and numbers in the different species. Several of the semi- and fully automated counters designed for use with human blood cannot be adjusted to give satisfactory analyses with samples from laboratory animals, or may be suitable for some species but cannot be used for leukocyte differential counts or platelet counts.

There are various publications that describe general requirements for hematology analyzers, and these are listed in the references as International Recommendations at the end of this chapter.

Several evaluations of hematology analyzers have been published for laboratory animals, and these publications include Weiser (1983, 1987a, 1987b), Tvedten and Wilkins (1988), Davies and Fisher (1991), Evans et al. (1991), Tvedten (1994), Bienzle et al. (1994), Byrne et al. (1994), Mische et al. (1995), Hagbloom et al. (1996), Pastor et al. (1997), Tabata et al. (1998), and Suzuki and Eguchi (1999).

Various technical issues have arisen with the development of different analyzers; these include:

1. The lysis of red blood cells, which then permits leukocyte counts. These lytic processes are dependent on mixing cells with reagents of chosen osmolalities. The osmotic fragility of erythrocytes varies between species: the greater the mean cell volume, the greater the osmotic resistance, or in other words, smaller erythrocytes are more susceptible to cell lysis (Perk et al., 1964; Coldman et al., 1969). The complete lysis of erythrocytes is essential if inaccurate counting of leukocytes is to be avoided. Several of the reagents used are described as isotonic, but this description applies to human cells and not for all animal species.

2. For hemoglobin determinations, there is an increasing usage of the anionic surfactant sodium dodecyl sulfate (SLS) to form a hemoglobin derivative after lysis of the erythrocytes. This is an alternative to the traditional cyanmethemoglobin method, and it is now favored for reasons of safety as a replacement of the cyanide required for the cyanmethemoglobin method. Erythrocyte lysis and reactions with SLS may differ between species (Evans and Smith, 1992).

3. Several analyzers use both an impedance and optical or laser methods for cell counting (denoted I for impedance and O for optical). Wide discrepancies between impedance and optical values must be investigated. Some of these differences may be due to platelet clumps, nonnucleated red blood cells, or red blood cells that are resistant to cell lysis.

4. Potential interferences that can cause errors in automated cell counting include aggregated platelets, nonnucleated red blood cells, Heinz bodies, agglutinated red cell agglutinins, and very high numbers of leukemic cells. Where there are very large numbers of leukocytes, the discrimination between leukocyte subpopulations may become blurred, and appropriate dilution may provide a more accurate differential count.

5. With species where red cells have lower mean corpuscular cell volumes, the discrimination between the distribution curves obtained for platelets and red cells may not be adequate with some analyzers for cell counting. In some instances, large platelets may be counted as red blood cells.

6. Some cell shapes are altered prior to counting, e.g., discoid to spheres, and this assumes that all cells will exhibit the same deformability with different laboratory species. Cells with a higher mean cell hemoglobin or increased viscosity may not deform completely; thus, the cell volumes are overestimated in some analyzers.

7. Hematocrit, mean cell volume, and mean cell hemoglobin are determined in different ways by the various analyzers. In some analyzers, hematocrit values are calculated from measured red blood cell count and mean corpuscular cell values, while other analyzers determine mean corpuscular volumes from red cell counts and measured hematocrit values. Hematocrit measurements performed by centrifugation techniques appear to differ between species, and centrifugal forces must be varied to obtain accurate measurements (Chien et al., 1965).

8. Differential leukocyte counting may use a cytochemical method based on the cell peroxidase activity, as the lymphocytes have very little activity in contrast to eosinophils, which show the strongest peroxidase activity in most species. This technique is used in conjunction with assessment of cell size and a second technique, where the cellular cytoplasm is stripped away and allows nuclei to be counted and cells to be classified as polymorphonuclear or mononuclear. Basophils retain their cytoplasm in this second process, although in some species basophils are overestimated because their cells do not lose their cytoplasm. Marmosets have relatively low eosinophilic peroxidase activity, and these cells may not be satisfactorily counted. These cytochemical techniques produce a sixth component of the differential counts, known as LUCs or large unstained/unclassified cells. These cells may include large activated lymphocytes (Zelmanovic et al., 1993).

9. Optical interference. Some xenobiotics or their metabolites may absorb light or fluoresce at similar wavelengths to those being used for measurement, and these compounds may cause interference. These interferences may be checked by analyzing solutions of the parent compound or metabolites, where these are available.

10. Interference due to hemolysis. Drug-induced hemolysis can be a key finding in toxicological studies, but hemolysis may be caused by procedures during sample collection or transportation. Hemolysis may interfere with chromogenic methods and coagulation studies (O'Neill and Feldman, 1989).

11. Interference due to lipemia. Erroneous high values for hemoglobin and red cell indices may be caused by hyperlipidemia. Interference by lipemia may be due to turbidity or light-scattering effects, or simply to plasma volume displacement due to high lipid levels (Gagne et al., 1977; Sharma et al., 1985; Canetero et al., 1996). Removal of the lipemic plasma after centrifugation and replacement with isotonic saline may confirm interference due to lipids. Fine lipid droplets also may be counted as platelets by some analyzers. Lipemia may be observed in animals receiving intravenous administration of fat emulsions, in genetically inherited hyperlipidemic strains, e.g., Wanatabe rabbits, or following major perturbations of lipid metabolism.

12. Blood substitutes. Pseudohemolysis is caused by red colorations imparted to plasma or serum by some blood substitutes (Chang, 1998). Other interferences, such as falsely elevated platelet counts, also have been reported (Cuignet et al., 2000).

13. Dilutions. For very small animals, e.g., transgenic mice, the blood volumes available are small, and some investigators use 1:2 or 1:3 dilutions to obtain results. These dilution procedures should be validated by the laboratory, as they are often less precise than the methods recommended by the manufacturer, and values may be lower than the linearity ranges determined for the analyzer.

14. Flagged data. Automated analyzers generate data flags that are based on preselected criteria, e.g., flagging values outside the reference range or where cell morphology differs from the criteria preselected. Blood films should be examined for confirmation of the apparent changes, and where the analyzer produces cytograms, these should be examined particularly where data are flagged. Flags may be described as distributional, e.g., when flagging outside reference limits, or morphological, where flags are due to change of cell shape or the distribution curve of cell populations.

Differences between analyzers have been reported for rat blood exchanged and analyzed in a multilaboratory study (Matsuzawa et al., 1996), and some differences are commonly seen when evaluating new analyzers in house (see previous references for analyzer evaluations). In an unusual example, when results obtained with different analyzers generally appeared to be similar, subsequent dosing of rats with 2-butoxyethanol showed that mean cell volume and hematocrit changes were detected by one analyzer but not another (Ghanayem et al., 1990). Different analyzer technologies rarely are the cause for marked differences between two laboratories evaluating the same compound.

Automated analyzers now offer a wide range of additional measurements, but many of these remain to be validated for animal species. These measurement include platelet crit, hemoglobin distribution width, platelet volume distribution width, immature granulocytes, etc.

BLOOD FILMS (SMEARS)

For some species where appropriate software is not available or validated, the differential count must be performed manually using blood films. Examination of these films also can provide useful information that adds to the primary cell measurements made using automated analyzer techniques, and can confirm or refute apparently erroneous or aberrant results. These films may be stained with modified reagents based on Romanowsky stains (May-Grünwald, Giemsa, and Wright). The differences between species when these general stains are used are slight, although more marked species differences occur with other leukocyte stains, which are less commonly used (Futamura et al., 1991). A bluish background is seen with heparinized samples that is not seen to the same extent with EDTA samples.

Problems associated with blood film preparation include irregular distribution of cells with thick borders and longitudinal patches in the film, and crenation due to delayed preparation or warming. There is often distinct observer bias when reading blood films, which can be reduced by exchanging slides between observers as part of an education program. For toxicology studies, all films should be read by one observer to minimize bias, and reviewed by a second observer if necessary.

HEINZ BODIES

The presence of Heinz bodies may be confirmed by staining blood films with supravital stains—brilliant cresyl blue, new methylene blue, or crystal violet. Heinz bodies may interfere with leukocyte counts on some analyzers, and hemoglobin measurements where there are large numbers of Heinz bodies (Beutler et al., 1955; Boelsterli et al., 1983).

RETICULOCYTES

Reticulocyte enumeration in peripheral blood provides information on hemopoiesis, and the reticulocytes can be counted manually or by automated counters. Reticulocytes can be counted manually by preparing blood films and staining the films with supravital stains—new methylene blue or brilliant cresyl blue—which stain the reticulum of precipitated ribonucleic acid (RNA), mitochondria, and organelles of the reticulocyte.

As reticulocytes may form only 1–5% of the red blood cell count, manual counts have inherent problems of imprecision (Evans et al., 1993).

Methods and instrumentation are now available for the automated counting of reticulocytes in several species, using flow cytometers or integrating them in the larger analyzer. These analyzers use differing dyes, which include thiazole orange, CD4K530, oxazine 750, acridine orange, and auramine O (Evans et al., 1993; Evans and Fagg, 1994; Collingwood and Evans, 1995; Perkins and Grindem, 1995; Riley et al., 2002). As with automated hematology counters, different reticulocyte counters appear to give differing results (Buttarello et al., 2001). With some automated techniques, some leukocytes appear to react with auramine O, and this leads to an erroneous count in humans.

Automated counting procedures allow the detection of much smaller changes of reticulocyte counts and improve the detection of some anemias. The ability to discriminate between reticulocyte subpopulations by their fluorescence, which reflects reticulocyte maturity, remains to be to thoroughly evaluated and exploited for laboratory animals (Houwen, 1992; Nusbaum, 1997; Riley et al., 2001).

METHEMOGLOBIN AND SULFHEMOGLOBIN MEASUREMENTS

Methemoglobin may be measured using co-oximeters. These analyzers measure absorption curves of various hemoglobin derivatives at multiple wavelengths, and then calculate the concentration of methemoglobin using various algorithms. Some oximeters have a limited number of algorithms for specific animal species, but the presence of sulfhemoglobin may cause inaccuracies due to spectral interference. Samples for methemoglobin should be sampled within 1 hour of collection, due to a reverse of its formation catalyzed by the methemoglobin reductase activity of the erythrocytes. Manual methods for methemoglobin and sulhemoglobin are time consuming and require the use of cyanide-containing reagents (Evelyn and Malloy, 1938).

COAGULATION

As discussed in Chapters 8 and 9, sample collection procedures can markedly affect coagulation measurements in smaller laboratory animals. Manual methods for performing assays of coagulation are suitable for diagnostic laboratories, but they are not sufficiently precise for toxicology studies, and therefore automated or semiautomated methods are preferred. Automated and semiautomated coagulometers/coagulyzers use several differing technologies to measure coagulation factors that in laboratory animals often differ in proportion from those found in humans. Differences due to reagents or analyzers have been described for prothrombin and activated partial thromboplastin time, and these differences in analytical systems are also recognized in human medicine (see Chapter 8). Some coagulyzers are unsuitable for laboratory animals because they cannot be adjusted to cope with the differing measurement times encountered, e.g., beagles and marmosets have much shorter prothrombin times—approximately 7 seconds, compared to approximately 14 seconds in man.

PLATELET AGGREGATION STUDIES

These methods are discussed in Chapter 8.

CHROMOGENIC FACTOR ASSAYS

Several coagulation factors, for example, plasminogen, plasminogen activator inhibitor, and tissue plasminogen activator, can be measured using chromogenic substrates, but such methods must be shown to be appropriate for each of the selected species. Apparent homology across species may not deliver reliable reactivity, or there may be no cross-reactivity.

GENERAL METHOD PRINCIPLES

All methods should be assessed for imprecision, linearity, and accuracy, and the related calibration and quality control of methods should be included in the laboratory operating procedures. There are several international guidelines for analyzer evaluations, and some of these appear in the references for this chapter and are listed as International Guidelines.

Imprecision of a method (more widely described as precision) is described by a numerical value (often expressed as standard deviation [S.D.] or coefficient of variation [CV]) obtained from a series of replicate measurements.

When interpreting counts determined by manual or automated counts, it is important to recognize the relative imprecision for each of the cell populations. As the number of cells within a cell population decreases, the imprecision of the measurement increases even with automated analyzers, where the number of cells counted is much greater than the 100, 200, or 400 cells counted manually. Using a manual 200-leukocyte count, and results expressed as percent, reported results may vary for 5% (from 2 to 10%), for 10% (from 6 to 16%) and for 25% (from 19 to 32%), where the values in parentheses indicate 95% confidence limits around those mean values (Rümke, 1960; Koepke, 1977).

For smaller laboratory animals, a pooled sample of blood must be used to provide sufficient volumes of blood to measure imprecision. Very rarely when preparing these pools, the erythrocytes may clump together and cause problems when performing these measurements.

Inaccuracy (more widely described as accuracy) describes the numerical value of the difference between the means of series of replicate values and the true value. Sometimes this difference is referred to as bias or systematic error.

Inaccuracy remains a major challenge for hematological measurements with laboratory animal samples in the absence of any primary reference material with accepted known values for individual species. Given that various technologies used for hematological measurements are known to result in differences with human and quality control samples, laboratories must ensure that the analyzer is calibrated and performing to the manufacturer's specifications. Comparison to other previous technologies and manual methods helps in determining relative inaccuracy or bias of a new method with a previous method, but with the knowledge that manual methods are less precise and there is no certainty that previous methods are accurate. Wherever possible, analytical methods should not be changed during a study (although this is sometimes a problem with studies of 2 years' duration). By ensuring that calibration and quality control procedures are followed, relative inaccuracy within a study and between studies for an assay should be maintained and minimized.

For coagulation studies, liquid or freeze-dried plasmas are commercially available with target values and ranges for most of the coagulation factors. These values are dependent on the analyzer and reagent systems employed. Some investigators use a pooled sample from healthy animals, which is assigned a theoretical value of 100%, and then express the results as a percent relative to this standard, but results can then vary with each pooled reference sample between studies, and the pool may not be stable during storage.

LINEARITY

This can be checked by preparing and analyzing serial dilutions of whole blood (or citrated plasma for coagulation assays), and comparing the determined values with the theoretical values calculated for the dilutions. For higher values, it is necessary to concentrate whole blood by centrifugation, but as the number of cells increases, the potential for interferences between cell populations increases, e.g., high leukocyte counts may appear as increased red cell counts, and high hematocrit values may affect hemoglobin measurements in some analyzers. There are commercially available materials for demonstrating linearity, but again, these are designed for use in human medicine. The options of using separate cell component fractions are generally not available, unlike in human medicine, where blood bank fractions may be used, e.g., platelets. The serial dilutions used for linearity checks can also help establish the analytical sensitivity, i.e., the minimal detectable change from one concentration to another.

Most manufacturers provide some indication of their analyzers' performance in terms of imprecision, linearity, and reportable ranges.

CALIBRATION AND QUALITY CONTROL MATERIALS

All automated counters should be calibrated correctly to obtain accurate results in animals using whole blood calibrants or the manufacturer's recommended methods. The calibration and quality control whole blood materials used in hematology are of animal origin, but these materials contain artificially fixed cells and do not currently use blood from laboratory animal species relevant for toxicology. Many hematology analyzers rely on single-point calibration at values that are normal for humans but differ from those of laboratory animals, and reliance on single-point calibration requires the analyzer to be linear over a wide range, which includes normal and extremely abnormal values as far as practicable.

To ensure correct calibration and consistent analyzer performance, quality control materials should be used for every batch of analyses performed. Many laboratories use trilevel materials similar in properties to the calibration materials, and with quoted assay values that are relative to low, normal, and high human blood samples. Whole blood materials obtained commercially for calibration and quality control purposes have a storage life of about 60 to 90 days when unopened and stored at 4 degrees Centigrade. The shelf life of these control materials reduces to about 5 days when a vial of whole blood material is opened.

For coagulation studies, liquid or freeze-dried plasmas can be used. The lyophilized coagulation controls and calibrants have longer unopened shelf lives, but they should be prepared as liquids and used on the same day as the analysis. It is good practice to cross-reference new batches of quality control materials with previous batches, as there may be slight differences. Sometimes these differences are unacceptable and are due to shipment and delivery problems.

Many of the current major analyzers or host computers have most, if not all, of the common tools used for quality control procedures, where out of range can be flagged and trends plotted, e.g., Levey–Jenning plots, Cusum charts, and application of Westgard rules (Westgard and Groth, 1981; Hinckley, 1997). Monitoring of internal quality control materials aids the recognition of drifts in accuracy, poor precision, and analyzer malfunctions. However, even the best quality control procedures will not detect random errors that may occur due to poor sampling or reagent problems. Gross errors unfortunately do occur in even the best managed laboratories—examples of these errors include incorrect sample labeling and wrongly identifying samples at the time of collection and analysis. These errors are often difficult to detect.

To aid control procedures, many laboratories participate in external quality assessment schemes (also termed external proficiency testing). In these assessment schemes the blood is usually manufactured from animals and fixed prior to distribution to a number of laboratories for comparative purposes. It is perhaps not surprising, given the various analyzer technologies, that results differ.

ELISA METHODS

For some hematologists, enzyme-linked immunoassays (ELISAs) represent newer technology, but these assays are increasingly used for coagulation tests and other tests such as erythropoietin. Often these assays are less precise than some routine clinical chemistry methods. Imprecision may be due to incomplete washing and aspiration of the wells, inadequate mixing of reagents within the wells, pipetting errors and contamination during the washing procedures, and incorrect timings for the various steps involving reagent additions, washing, and incubation. In addition to these problems, the measuring range may be inappropriate for the species, or there may be a lack of immunospecificity.

MAINTENANCE

Based on manufacturers' recommendations and experience, laboratories must perform preventative maintenance procedures on the analyzers being used for measurements. Some additional maintenance procedures will be implemented in response to internal quality control performance or analyzer malfunctions. Calibration and quality control performances should always be checked after maintenance visits, as replacement parts, disturbances of the optics, and hydraulics sometimes affect subsequent analyzer performance and cause unsatisfactory performance (described as postmaintenance visit trauma [PMT]). These maintenance procedures form part of good laboratory practice (Weinberg, 1995; FDA, 2007).

BIOHAZARDS AND CHEMICAL SAFETY

When transporting, receiving, or analyzing diverse blood samples, every effort should be made to reduce the chemical and biohazards for laboratory staff (Truchard et al., 1994; WHO, 2004). Suitable containers/bags should be used to transport samples from the animal care buildings to the laboratory. Allergies to laboratory animals remain a risk for laboratory workers, and efforts should be made to minimize exposure to animal dander, etc. Although most laboratories use purpose-bred animals, occasionally bacteria, parasites, and yeasts may be found in blood (Owen, 1992). Where immunosuppressants or immunostimulants are being tested, this may increase the risks of viral infections, and potential viral transmission, and where wild-caught nonhuman primates or other species are being studied, for example, in Third World countries, the risks from viral infections must be considered. Guidance for periodic cleaning and disinfection of analyzers is usually provided by manufacturers, but more frequent measures may be required following analysis of infectious materials.

During the analytical preparation stages, there are several risks for accidents that should be minimized. These include aerosols produced in breakages by centrifugation, glass breakages of slides or microhematocrit tubes, sharp needles, and metal foil caps.

Adequate warnings are usually provided by analyzer manufacturers on electrical and laser beam hazards. Chemical hazards are usually indicated by the reagent manufacturers, and they include warnings relevant to the use of cyanide for hemoglobin determinations and disinfection solutions. Some reagents used for flow cytometry and reticulocyte counting are potentially carcinogenic and should be handled accordingly to minimize the risks. Samples from some studies may contain radioactivity, and these should be treated according to local rules pertaining to radiation safety.

REFERENCES

INTERNATIONAL RECOMMENDATIONS

ICSH. 1993. ICSH recommendations for the measurement of erythrocyte sedimentation rate. *J. Clin. Pathol.* 46:198–203.
ICSH. 1994. Guidelines for the evaluation of blood cell analysers including those used for differential leucocyte and reticulocyte counting and cell marker applications. *Clin. Lab. Haematol.* 16:157–74.
ICSH. 1998. Proposed reference method for reticulocyte counting based on the determination of the reticulocyte to red cell ratio. *Clin. Lab. Haematol.* 20:77–79.
ICSH. 2001. RBC/platelet ratio method: A reference method. *Am. J. Clin. Pathol.* 115:460–64.
NCCLS. 1992. Reference leukocyte differential (proportional) and evaluation of instrumental methods: Approved standard. Publication H20-A. *Clin. Lab. Haematol.* 14:69–84.
NCCLS. 1997. *Methods for reticulocyte counting (flow cytometry and supravital dyes). Approved guideline.* Publication H44-A. Wayne: NCCLS.
NCCLS. 2001. *Evaluation of the linearity of quantitative analytical methods: Proposed guideline.* Publication EP6-P2, 2nd ed. Villanova: NCCLS.

HEMATOLOGY ANALYZERS

Groner, W., and Simson, E. 1995. *Practical guide to modern hematology analyzers.* New York: John Wiley & Sons.
Lombarts, A. J. P. F., Koevert, A. L., and Leijnse, B. 1986. Basic principles and problems of haemocytometry. *Ann. Clin. Biochem.* 23:390–44.
Paterakis, G. S., Laoutaris, N. P., Alexia, S. V., Siourounis, P. V., Stamulakatou, A. K., Premetis, E. E., Sakellarious, Ch., Terzoglou, G. N., Papassotiriou, I. G., and Loukopoulos, D. 1994. The effect of red cell shape on the measurement of red cell volume. A proposed method for the comparative assessment of this effect among various haematology analysers. *Clin. Lab. Haematol.* 16:235–45.
Schoentag, R. A. 1988. Hematology analyzers. *Clin. Lab. Med.* 8:653–73.

ANIMAL HEMATOLOGY ANALYZER EVALUATIONS

Bienzle, D., Jacobs, R. M., Lumsden, J. H., Grift, E., and Tarasov, L. T. 1994. Comparison of two automated multichannel haematology analysers in domestic animals. *Comp. Haematol. Int.* 4:162–66.
Byrne, R. F., Andrews, C. M., Libretto, S. E., and Mifsud, C. V. 1994. Canine and feline haematology analysis: Comparative performance of Technicon H*1 and AVL MS8 VET analysers. *Comp. Haematol. Int.* 4:212–17.

Davies, D. T., and Fisher, G. V. 1991. The validation and application of the Technicon H*1 for the complete automated evaluation of laboratory animal haematology. *Comp. Haematol. Int.* 1:91–105.

Evans, G. O., Fagg, R., Flynn, R. M., and Smith, D. E. C. 1991. Haematological measurements in healthy beagle dogs using a Sysmex E-5000. *Sysmex J.* 14:127–28.

Hagbloom, R., Begg, K., Mast, B., Stewart, N., Houwen, B., and Culp, N. 1996. Performance evaluation of the K-4500™ using animal blood. *Sysmex J.* 6:147–54.

Mische, R., Deniz, A., and Weiss, J. 1995. Untersuchung zur automatischen Zellzählung aus Katzenblut. *Dtschtierarztl. Wschr.* 102:435–40.

Pastor, J., Cuenca, R., Verlarde, R., Vinas, L., and Lavin, S. 1997. Evaluation of a hematology analyser with canine and feline blood. *Vet. Clin. Pathol.* 26:138–47.

Suzuki, S., and Eguchi, N. 1999. Leucocyte differential analysis in multiple laboratory species by a laser multi-angle polarized light scattering separation method. *Exp. Anim.* 48:107–14.

Tabata, H., Kubo, M., Izumisawa, S., Narita, K., Matsuzawa, T., and Sakai, T. 1998. A parallel comparison of three automated multichannel blood cell counting systems for analysis of blood of laboratory animals. *Comp. Haematol. Int.* 8:66–71.

Tvedten, H. 1994. Canine automated differential leukocyte count using a hematology analyzer system. *Vet. Clin. Pathol.* 23:90–97.

Tvedten, H. W., and Wilkins, R. J. 1988. Automated blood cell counting systems: A comparison of the Coulter S-Plus IV, Ortho ELT-8/DS, Ortho ELT-8/WS, Technicon H-1 and Sysmex E5000. *Vet. Clin. Pathol.* 17:47–54.

Weiser, M. G. 1983. Comparison of two automated multi-channel blood cell counting systems for analysis of blood in common domestic animals. *Vet. Clin. Pathol.* 12:25–32.

Weiser, M. G. 1987a. Modification and evaluation of a multichannel blood counting system for blood analysis in veterinary haematology. *J. Am. Vet. Assoc.* 190:411–15.

Weiser, M. G. 1987b. Size referenced electronic leukocyte counting threshold and lysed leukocyte distribution of common domestic animal species. *Vet. Pathol.* 24:560–63.

ISSUES ASSOCIATED WITH HEMATOLOGY ANALYZERS

Osmolality

Coldman, M. F., Gent, M., and Good, W. 1969. The osmotic fragility of mammalian erythrocytes in hypotonic solutions of sodium chloride. *Comp. Biochem. Physiol.* 31:605–9.

Perk, K., Frei, Y. F., and Herz, A. 1964. Osmotic fragility of red blood cells of young and mature domestic and laboratory animals. *Am. J. Vet. Res.* 25:1241–48.

Hemoglobin with SLS

Evans, G. O., and Smith, D. E. C. 1992. Preliminary studies with an SLS method for haemoglobin determination in three species. *Comp. Haematol. Int.* 2:101–2.

Hematocrit

Chien, S., Dellenback, J., Usami, S., and Gregersen, M. I. 1965. Plasma trapping in hematocrit determination. Differences among animal species. *Proc. Soc. Exp. Med.* 119:1155–58.

Cytochemical Leukocyte Counting

Zelmanovic, D., Garcia, R. A., Leitner, T., and Jones, V. 1993. H1*E multi-species haematology. *Miles Sci. J.* 15:7–13.

Hemolysis Interference

O'Neill, S. L., and Feldman, B. F. 1989. Hemolysis as a factor in clinical chemistry and hematology of the dog. *Vet. Clin. Pathol.* 18:58–68.

Lipid Interference

Canetero, M., Conejo, J. R., and Jiménez, A. 1996. Interference from lipemia in cell count by hematology analyzers. *Clin. Chem.* 42:987–88.

Gagne, C., Auger, P. L., Moorjani, S., Brun, D., and Lupien, P. J. 1977. The effect of hyperchylomicronemia on the measurement of haemoglobin. *Am. J. Clin. Pathol.* 68:584–86.

Sharma, A., Artiss, J. D., Stranbergh, D. R., and Zak, B. 1985. The turbid specimen as an analytical medium: Hemoglobin determination as a model. *Clin. Chim. Acta* 147:7–14.

Blood Substitutes

Chang, T. M. S., ed. 1998. *Blood substitutes: Principles, methods, products and clinical trials.* Vols. 1 and 2. Basel: Karger AB.

Cuignet, O. Y., Wood, B. L., Chandler, W. L., and Spiess, B. D. 2000. A second generation blood substitute (perfluorodichorooctane emulsion) generates spurious elevations in platelet counts from automated hematology analyzers. *Anesthesia Analgesia* 90:517–22.

COMPARISON OF HEMATOLOGY ANALYZERS

The references listed above for hematology analyzers also include some comparative data.

Ghanayem, B. I., Ward, S. M., Blair, P. C., and Mathews, H. B. 1990. Comparison of the hematologic effects of 2-butoxyethanol using two types of hematology analyzers. *Toxicol. Appl. Pharmacol.* 106:341–45.

Matsuzawa, T., et al. 1996. A survey of the result of haematological parameters, using a common rat blood sample in Japanese laboratories. *Comp. Haematol. Int.* 6:125–33.

LEUKOCYTE COUNTING

Futamura, Y., Matsumoto, K., and Furaya, T. 1991. Characteristics of staining of leukocytes in experimental animals. *Exp. Anim.* 40:121–25.

Koepkc, J. A. 1977. A delineation of performance criteria for the determination of leukocytes. *Am. J. Clin. Pathol.* 68(Suppl.):202–6.

Rümke, C. L. 1960. Variability of results in differential counts in blood smears. *Triangle* 4:154–58.

HEINZ BODIES

Beutler, E., Dern, R. J., and Alving, A. S. 1955. The hemolytic effect of primaquin. VI. An in vitro test for sensitivity of erythrocytes to primaquine. *J. Lab. Clin. Med.* 45:40–50.

Boelsterli, U. A., Shie, K. P., Brandle, E., and Zbinden, G. 1983. Toxicological screening models: Drug-induced oxidative hemolysis. *Toxicol. Lett.* 15:153–58.

Reticulocyte Counting

Buttarello, M., Bulian, P., Farina, G., Temporin, V., Toffolo, L., Trabuio, E., and Rizzotti, P. 2001. Flow cytometric reticulocyte counting. Flow cytometric evaluation of five fully automated analyzers: An NCCLS/ICSH approach. *Am. J. Clin. Pathol.* 115:100–11.

Collingwood, N. D., and Evans, G. O. 1995. Reticulocyte counts in laboratory animals using the R2000 flow cytometer. *Sysmex Int. J.* 5:126–29.

Evans, G. O., and Fagg, R. 1994. Reticulocyte counts in canine and rat blood made by flow cytometry. *J. Comp. Pathol.* 111:107–11.

Evans, G. O., Fagg, R., and Smith, D. E. C. 1993. Manual absolute reticulocyte counts in healthy Wistar rats. *Comp. Haematol. Int.* 3:116–18.

Houwen, B. 1992. Reticulocyte maturation. *Blood Cells* 18:167–86.

Nusbaum, N. J. 1997. Red cell age by flow cytometry. *Med. Hypotheses* 48:469–72.

Perkins, P. C., and Grindem, C. B. 1995. Evaluation of six cytometric methods for reticulocyte enumeration and differentiation in the cat. *Vet. Clin. Pathol.* 24:37–43.

Riley, R. S., Ben-Ezra, J. M., Goel, R., and Tidwell, A. 2001. Reticulocytes and reticulocyte enumeration. *J. Clin. Lab. Anal.* 15:267–94.

Riley, R. S., Ben-Ezra, J. M., Tidwell, A., and Romagnoli, G. 2002. Reticulocyte analysis by flow cytometry and other techniques. *Hematol. Oncol. Clin. N. Am.* 16:373–420.

Methemoglobin and Sulfhemoglobin

Evelyn, K. A., and Malloy, H. T. 1938. Microdetermination of oxyhemoglobin, methemoglobin and sulfhemoglobin in a single sample of blood. *J. Biol. Chem.* 126:655–62.

Quality Control

Hinckley, M. C. 1997. Defining the best quality-control systems by design and inspection. *Clin. Chem.* 43:873–79.

Westgard, J. O., and Groth, T. 1981. Design and evaluation of statistical control processes: Application of a computer "quality control simulator" program. *Clin. Chem.* 27:1536–47.

Good Laboratory Practice

FDA. 2007. *Good laboratory practice for non-clinical laboratory studies.* Title 21 CFR 58, electronic code of Federal Regulations. Food and Drug Administration.

Weinberg, S., ed. 1995. *Good laboratory practice regulations.* New York: Marcel Dekker.

Safety

Owen, D. G. 1992. Parasites of laboratory animals. *Laboratory animal handbook*, Vol 12. London: Royal Society of Medicine Press.

Truchaud, A., Schnipelsky, P., Pardue, H. L., Place, J., and Ozawa, K. 1994. Increasing the biosafety of analytical systems in the clinical laboratory. *Clin. Chim. Acta* 226:S5–13.

WHO. 2004. *Laboratory biosafety manual.* World Health Organization.

11 Data Processing and Interpretation

Having produced the data, the next steps are to analyze the data and summarize the observed effects using experience and statistics where appropriate.

DATA DEFINITIONS AND SECURITY

The laboratory's standard operating procedures should ensure that data in paper or electronic format are stored securely. Editing of data must be properly tracked with reasons for changes and must identify the person making data changes, and the person changing data must be authorized to make such changes. The constitution of the raw data must be described, as these data must be retained and eventually archived for regulated studies. Some of the definitions and requirements for electronic raw data are outlined by a FDA guidance document (Food and Drug Administration, 2003).

Each regulatory study design/protocol/plan defines the tests required, and any additional tests must be approved by formal approved amendments to the study plan. Hematological analyses may present some additional problems that have to be considered when defining the raw data. Many hematology analyzers measure more parameters than those required by the study plan, and it is not possible to prevent these results by reconfiguring the analyzer. These extra data are redundant, and some can be simply dismissed, as their value has not been evaluated thoroughly for animal samples. If these extra data are not to be stored on a host computer, then the unwanted data can be suppressed from the data transmissions from the analyzer, but these data manipulations should be described in a standard operating procedure.

For the cytogram data produced by a hematology analyzer, the laboratory needs to consider the definition of these data. Traditionally these hematology cytogram data sets were not stored except as temporary electronic files on the analyzer, and the stored data were numeric, e.g., lymphocyte counts stored as numeric counts and not as the cytogram. The cytograms produced by hematology analyzers or flow cytometers may be defined as raw data and require storage in either paper or electronic formats. For the majority of hematology analyzers, the software gating programs for separate cell populations used in the cytograms are fixed or locked, but in flow cytometry cell population gating is often flexible and dependent on the analyst. The stored data format of cytograms and population gating records is therefore important. Where the cytograms are printed, the hardcopy can be defined as the raw data; when storing electronic cytogram records, the data often cannot be interpreted by the next generation of analyzers, or requires a complex software program to retrieve the data, and this is a major limitation.

REPEATED ANALYSES ON THE SAME SAMPLE

Unlike toxicokinetics where the samples may be analyzed in triplicate or quadruplicate, most hematology samples for core tests are analyzed once due to the small sample volumes available. Trained hematologists will invariably spot aberrant sample values that have been affected by poor sample aspiration, small clots, or unusual analyzer flags and cytograms when using an automated counter. In some instances it may be that a sample was not completely mixed or the hydraulic sampling did not work correctly due to a slight blockage. Examination of the analyzer cytograms is important, and total reliance should not be placed on the flags generated by the analyzer. A sample for coagulation studies may not clot or may be partially clotted prior to analysis, and this result may be artifactually incorrect. The natural responses to such findings are to repeat the analysis for confirmation where there is sufficient sample, and to examine the blood film.

The laboratory should examine the reasons for repeating sample analysis and why certain results are accepted or rejected. If a value is outside the flagged ranges preset by the analyzer, why should you repeat one sample if all the surrounding sample values appear to be within these ranges, and having repeated the analysis, is the first or second result correct? If the number of replicate measurements varies between individual animals, then the inclusion of all data can be statistically complex, requiring decisions for inclusion and exclusion of data. Whatever the final decision as to the validity of results, the laboratory should ensure that the standard operating procedures provide sufficient adequate guidance for repeating analyses.

INTRA- AND INTER-INDIVIDUAL VARIATIONS

In the two previous chapters, preanalytical and analytical variables have been considered, and these can be described as the biological variance and the analytical variance, respectively. The total variance is the sum of the biological and analytical variances, with a third component added when two or more studies are compared, for no two studies are identical in every respect.

Biological variation occurs within (intra-) and between (inter-) animals. The animals used in toxicity studies are purposely bred with known genetic backgrounds using a carefully controlled environment with special reference to pathogens and adequate diet. The exceptions are for some nonhuman primates where the genetic backgrounds and refinements are less well known in the captive-bred populations. In some countries the supply and control of laboratory animals is less consistent, and the laboratory animal populations are more variable. For many years laboratory animal breeders worked to produce more uniform animal populations, but more recently there has been a counter-debate suggesting that the use of less well-controlled genetic populations would aid the detection of idiosyncratic adverse reactions.

A simple procedure to obtain an estimate of biological variation can be used in some larger species, for example, in dogs, where values for an individual animal can be compared for pretreatment and serial samples to obtain an indication of the intra-animal variation. These data can also be used for comparison with other animals within and between groups to obtain indications of the intra- and inter-animal variations of data. In the larger laboratory animals, where blood volumes permit, it is useful to take two pretreatment values; this has several advantages in that it gives a stronger indication of the intra- and inter-animal variations in studies where the group sizes are usually small, and also allows some confirmation of outlying values particular to an individual animal. This may enable the animal to be excluded from the study or reassigned to a different treatment group where the outlying values are unlikely to affect the overall conclusions of the study.

It is more difficult to obtain estimates of the biological variation for the smaller laboratory animals such as rodentia because the volumes of blood required for repetitive analysis result in changes that include body fluid balance perturbations, reticulocyte and bone marrow response to blood loss, and leukocytosis associated with stress of frequent blood collections. In general, the variations due to analytical components in hematological measurements usually are much less than variations due to preanalytical factors. Compared to analytical variance, the biological variance components are much greater in healthy rats (Weil, 1982; Gaylor et al., 1987; Carakostas and Banerjee, 1990) and healthy dogs (Jensen et al., 1998), and biological variations are higher than in humans because there are many more preanalytical effects that cannot be controlled (see Chapter 9). These inter- and intra-variances can be expected to be higher with toxicity associated with escalating doses.

CRITICAL DIFFERENCES

The calculation of critical differences is a tool for following changes in consecutive measurement in an individual animal. Critical differences can be calculated by the following formula (Fraser and Harris, 1989):

1.96 × the square root of 2 × (biological coefficient of variation^2 for intra-individual + analytical coefficient of variation^2)

For example, based on replicate measurements in healthy normal beagles, Jensen et al. (1994, 1998) calculated that the critical differences between two consecutive measurements were about 18% for red blood cell count, hematocrit, and hemoglobin, and 35% for total white blood cell counts at a significance level of 0.05.

In human medicine, there have been attempts to define critical values, which are values so different from the reference values that they represent a pathophysiological state that is life threatening unless some action is taken to rectify the problem. Examples of human critical values are hemoglobin values of less than 5 g/dl, platelet counts of less than 30×10^9/l, and prothrombin values greater than times 40. In toxicology, we can also readily identify such outlying or extreme values, and then need to examine the data for similarity between individual animals or trends toward such extreme values. Occasionally the test values may be similar to the critical values, but reflect a major problem in the sampling collection procedure. Remember, samples taken from animals *in extremis* often are widely divergent from reference values.

REFERENCE VALUES

Despite numerous attempts to persuade investigators from using the terminology, the term *normal range* still is frequently used rather than *reference values*. The reference (or normal) range values are determined for samples collected from healthy animals (reference individuals with described procedures for sample collections, sample processing, analytical methodology, etc.) (IFCC, 1987). The reference intervals, as often defined by the lower and upper reference limits, are calculated to include usually the central 95% of the population. The values outside the reference range may not be abnormal as implied by the use of the term *normal range*, but merely reflect that 5% of individual values in healthy animals lie outside the determined reference range.

Samples from a number of animals must be examined to determine the reference range, and this number should exceed 40, and ideally be at least more than 100 animals. When more than 120 animals are sampled to establish the reference values, there seems to be a diminishing improvement or alteration in the ranges obtained when these ranges are compared to ranges using up to 120 animals. For fewer than 40 animals, the lowest and highest observed values are often the best estimate of the central 95% reference interval.

Reference ranges should be qualified by:

Literature based or determined in-house
Strain, age, and gender
Dosing regime—oral, intravenous, or inhalation
Animal supplier
Sample collection method
Analytical methods
Exclusion criteria
Number of animals
Period of data collection

A 10-year-period database, for example, has limited use if the items listed above are not logged, as several of these variables will have changed during the 10-year period. Data obtained by new methodology, e.g., following the introduction of a new hematology analyzer, should be compared with current methodology to detect possible changes to existing reference ranges. Some investigators rely heavily on reference ranges, but accumulation of such data without the cognizance of the influence of preanalytical and analytical factors can lead to misleading interpretations. Reference ranges should not be used blindly to dismiss all values within the range as being of little or no consequence.

DATA DISTRIBUTIONS

The distribution of data should be examined in determining reference ranges and in toxicological studies. A normal (or Gaussian) distribution peaks in the middle, and is perfectly symmetrical or bell shaped about the middle of the distribution. Parametric testing assumes that the data are normally distributed, the distributions are similar in the groups being compared, and there are no extreme or outlying values. In most cases modest departures from these assumptions will allow satisfactory parametric analysis of the data. This parametric analysis may lead to the detection of smaller differences between the populations.

In many instances, hematological data are not normally distributed, and the distributions of data for control or test groups of rats in similar toxicological studies are not always identical (Weil, 1982). The distributions may be skewed to the left (negative in direction) or leptokuric (showing positive kurtosis). Nonparametric analyses should be used to estimate the 2.5 to 97.5 percentiles to define a 95% reference interval for non-Gaussian distributions. Sometimes transformation of data, e.g., by log transformation, can be used to change the data and make it more suitable for analysis. Such transformations can be helpful in establishing some references values.

If data are collected over a number of days or weeks, data can be inspected for any interbatch variations that may affect the final interpretation and processing, and any apparent outliers should be examined for any obvious confounding factors, e.g., extreme difficulties in obtaining blood samples or evidence of gross hemolysis, which would lead to exclusion of data items. Due to the instability of the samples for most blood measurements and the relatively small volumes available, the option to repeat sample analyses is rare if the investigator wishes to confirm outlying values.

STATISTICAL METHODS

The statement attributed to Benjamin Disraeli—"There are three kinds of lies: lies, damned lies, and statistics"—is unfortunately sometimes quoted to dismiss statistical findings that do not support conclusions based on the biological findings. Investigators are far more content when the statistical results support the conclusions based on the biological findings. Statistical methods are one of the tools to be used in the overall interpretation of a study but not the only tool, and graphical plots of data may be useful additional procedures. Making hematological measurements can produce vast amounts of data, but we need to consider some of the foundations used in toxicological experiments.

Some common terminology includes:

Confidence limits: Relate to a normal distribution, and commonly the limits applied are 95% or mean ± 2 S.D.

n **or** N: The number of samples included in a study group, or used in a single group for method comparisons.

Power: Relates to the probability of a statistical test rejecting the null hypothesis when it is false.

Probability values (p **or** P): Usually presented as $p < 0.001$ (***), $p < 0.01$ (**), $p < 0.05$ (*). (See the questioning of the overuse of $p < 0.05$ by Jones, 1988; Romano, 1988).

Standard deviation (S.D.): The square root of the variance and a measure of the scatter of values about the mean value. Smaller values are indicative of tighter clustering of data about the mean. By calculating the mean and S.D.

for a reference population, ±1 S.D. will contain 68% of all values, ±2 S.D. will contain 95.5% of all values, and ±3 S.D. will contain 99.7% of all values.

Standard error of the mean (S.E.M. or S.E.): The magnitude of S.E.M. values is dependent on the S.D. and number of results within the group. Although S.E.M. values appear smaller than S.D. values, there is doubtful value in expressing the results as S.E.M, and results should be expressed as S.D. in preference wherever possible.

Transformation of data: For distributions that are not parametric, data can be converted to $\log_{10}$ or $\log_e$ values to obtain a lognormal distribution prior to applying a parametric method of comparison, such as t-test or ANOVA.

CONTROL GROUPS

Control groups should always be included in study designs, and they should be gender and sex matched. Even if the test compound is designed for use with one sex, e.g., female hormone therapy, male animals also should be studied to identify toxicities unrelated to the pharmacological actions of the test compound. If the test vehicle is novel, then an additional control group should be included in the study design. For inhalation studies, it is useful to include both an air-only control group and a vehicle control group. Animals should be always be randomized when assigned to groups and for sampling procedures.

Usually a study design will include a control group that will allow a comparison between the test result variations that occur in the absence of the test compound. It is important to recognize that even within this control group, there may be one or more individuals that have some data that can be ascribed as abnormal or outside an expected range for healthy animals. In studies of dogs and nonhuman primates where the animal numbers may be less than five per group, it is useful to take at least two samples prior to the commencement of the studies about 1 week apart. This brings the benefits of having some indication of intra- and inter-animal variation and introduces naïve animals to the local blood collection procedures and animal care staff. These pretreatment results may on a few occasions affect the final allocation of animals to groups where there are outlying data that may have a direct consequence on the study, e.g., dogs with a factor VII deficiency or very low platelet count in a study involving novel antiplatelet aggregating agents.

The size of study should be determined by using power analysis to avoid excessively large studies and to minimize the number of animals in the study (Curtis et al., 1990; Desu, 1990; Mann et al., 1991; Festing, 1995; Festing et al., 2002). Unfortunately, this is not always the case, as studies tend to follow traditional in-house designs. The numbers of animals in each treatment group are usually larger in rodent studies than the group numbers for other species; this tends to produce more uniform results in rodent studies, but even here group sizes may be no more than twenty-five animals per group per sex. In preliminary rodent studies, the group sizes may be small, and in dog and nonhuman primate studies the group sizes may often be less than six per group. Studies are sometimes said to be "underpowered" in the numbers of animals within groups, i.e., the number of animals will not be sufficient to detect change. This is against a background of an increasing pressure to minimize the number of animals in studies. In designing studies there may be some other restrictions; for example, the amount of blood to be taken at necropsy or during a study where there may be demands to take samples primarily for pharmacokinetics in rodents may require the use of a separate group of animals, but this necessarily increases the total number of animals required.

STATISTICAL ANALYSES

Many laboratory information management systems (LIMSs) capture the data and contain simple programs that calculate group mean and standard deviation (or standard mean error) values— sometimes termed descriptive statistical data. The user should recognize that some of these statistical programs work with the assumptions that the group sizes are adequately sized and that all distributions are Gaussian. These assumptions may be incorrect; thus, all individual data should be examined and outliers identified, as these will affect both the mean and the standard deviation values. In small groups of fewer than ten animals, outliers may have a considerable effect on these calculated values, and it is important to identify these individual animals.

Although it is always the outliers that catch the reader's eye, it must always be remembered that there may be animals with values within the reference range that have or are undergoing a reaction to the compound. For example, a leukemic animal may show normal leukocyte differential percentage counts but have an abnormal white cell count, or depending on the timing of sample, the reticulocyte count may appear normal but there may be a reticulocyte response found earlier or later on other sampling days.

There are numerous statistical methods for parametric and nonparametric analyses (see Table 11.1), and the choice of methods is often made locally (Waner, 1992).

Some laboratories use decision trees in applying several of these tests sequentially, where the statistical analyses usually include an examination of data distributions, application of favored parametric and nonparametric tests, and a test for dose–response relationships (Morgan, 1996; Dickens and Robinson, 1996; Gad, 1998). The use of inappropriate statistical methods should be avoided, and their application is limited where the group sizes are small, e.g., in dog and nonhuman primate studies. Using a nonparametric method at one time point and a parametric method at another time point should also be avoided if possible, as the distributions of data may differ.

Comparison of the groups of dosed animals with the corresponding control group is one step, but a more useful test for the effect of the test compound is to examine the data for a dose-related effect, as many true effects of treatment tend to produce an increase or decrease of test values with increasing dosage. However, establishing that there is a trend effect with dosage does not imply that there is always a risk at the lowest dosage.

Multivariate analysis or principal component analysis (PCA) can be employed to explore relationships between a number of hematological variables; the technique can be used to create data clusters that can be used to demonstrate similarities between animals within a group or to identify outliers (Festing et al., 1984). This technique appears to be rarely used, but it may have application when linking the results of new and rarer tests with the core tests.

TABLE 11.1

Some Statistical Tests Applied in Toxicology Studies

Tests for Nonnormality

Kolmgorov–Smirnov

Chi-square

Shapiro–Wilk

Tests for Homogeneity of Variance

Bartlett's test

Levene's test

For Normally Distributed Data

Analysis of variance (ANOVA)

Analysis of covariance (ANCOVA)

Pearson's correlation coefficient

Linear regression to test for dose-effect trends

Pairwise comparison

Duncan's multiple test range

Dunnett's test

Williams' t-test

Student's t-test

Fisher's least significant difference (LSD) test

For Nonparametric Distributed Data

Kendall's coeffcient of rank correlation

Mann–Whitney U test (similar to t-test)

Wilcoxon signed rank test (paired or matched pairs)

Kruskal–Wallis ANOVA

Dunn's test

Shirley's test

Jonckheere's test

Several references for some statistical approaches are listed under the headings "Design and Size of Animal Studies" and "Statistical Approaches" at the end of this chapter.

Given the small group sizes, statistically significant differences often are found in the hematology data sets. For example, in a study of four dose groups per sex and with fifteen hematological measurements per sample, it is not uncommon to find at least one or two statistical differences between groups. These differences are highlighted by probability values (which may be indicated as *, **, ***), and this sometimes leads to "star gazing," where every attempt is made to give biological significance to a statistical probability value but where some of these differences are due to chance. There is common ground between biological findings and statistical significances where both show effects or both show no effects. It is more difficult when there is either a biological or statistical significance not supported by the other; these differences can sometime lead to data torture: "If you torture your data long enough, they will tell you what you want to hear" (Mills, 1993).

Data should be examined for dose relationships, and possible expected relationships between several parameters, e.g., alteration of erythrocytic parameters and reticulocyte counts. Usually an alteration of test results will be supported by other evidence gathered mainly from histopathology, clinical observations, and other clinical pathology disciplines that allow confirmation of a hematological finding and a decision as to whether an effect is due to toxicology, pharmacology, or is a biological variation. Sometimes the changes appear to be minor and can be dismissed due to a lack of corroborating evidence, but it is important to reflect on these findings in subsequent studies to see if a consistent pattern emerges.

Some questions for determining an effect on a hematological measurement are:

Due to pharmacological action?
    Normal or exaggerated?
    Desirable (expected) or adverse?
Dose related or a hormetic response?
Toxicity?
Due to test compound (or enantiomer)?
Due to metabolite?
Due to prodrug component rather than drug itself?
Due to vehicle or excipient?
Due to preanalytical factor?
Due to interference with analytical methodology?
Greater than inter- and intra-animal variation?
Greater than analytical variation?
Changes occur over time and in what sequence?
Related to other hematological measurements?
Occurs in one or more species?
Due to a combination of any of these factors?

It is also important to recognize that effects in two or more species may vary between the species, with perhaps no effect in one species, but an effect in another, or effects are opposite, with the changes being increased in one species and decreased in another.

SENSITIVITY AND SPECIFICITY

Sometimes with a new test, the question is asked as to how predictive or sensitive is the test. (Sensitivity of a test can have a very different meaning to a drug analyst looking for the lowest level of quantification.)

The concepts of sensitivity and specificity are applied to assays used for disease diagnosis. The sensitivity of an assay is the fraction of those with a specific disease that the assay correctly predicts, and specificity is the fraction of those without the disease that the assay correctly predicts, and these can be expressed in formulae where (Griner et al., 1981):

TP = true positive—number of affected individuals correctly classified by the test
FP = false positive—number of nonaffected individuals misclassified by the test
TN = true negative—number of nonaffected individuals correctly classified
    by the test
FN = false negative—number of individuals misclassified by the test

Then, sensitivity is expressed by

$$TP/(TP + FN) \times 100$$

and specificity is expressed by

$$TN/(FP + TN) \times 100$$

If a test has a high sensitivity with a value of 0.9999, then there is a 99.99% chance of the test being positive; conversely, if a test has a high specificity of 0.9999, then there is a 99.99% chance of the test being negative. If these values are applied to a population of 10,000 subjects where there is a 1.5% incidence rate of the condition being tested for, there will be 151 positive results where 1 result is false, and the chances of the test being truly positive are 150 out of 151, i.e., the test is highly predictive. However, if the incidence rate of the condition is 1 in 10,000, then one true false negative but two positive results—one truly positive and one falsely positive—can be expected, i.e., there is a fifty–fifty chance of the positive results being true.

These expressions for sensitivity and specificity can be used to assess the diagnostic value of a particular test, and to assess the predictive values of tests or groups of tests when extrapolating data from animals to human data (see Schein et al., 1970; Schein and Anderson, 1973).

EXTRAPOLATION OF ANIMAL DATA TO HUMANS

A toxicology study should be designed to determine the range of toxicities in the selected animal species, enable extrapolation to other species, including humans, and determine safe levels of exposure (Garratini, 1985; Zbinden, 1991).

When the concepts of sensitivity and specificity were applied to dog and monkey studies with anticancer drugs, it was found, not unexpectedly, that hematological measurements performed well in predicting anemias, leukopenias, and thrombocytopenias, but there were some failures and differences between dogs and monkeys (Schein et al., 1970; Schein and Anderson, 1973). Two later studies of more than eighty compounds given to rodents and nonrodents demonstrated that hematological effects were relatively common—approxinately 30% for rodent and dog studies—and there were species differences, with these findings being less common in monkeys (Heywood, 1981a, 1981b).

In a collective study, data for 150 varied compounds were collated and examined for concordance between toxicities found for animal and human data (Olson et al., 2000). Human hematological toxicities correlated to a high degree (90%) with animal findings for both nonrodent and rodent studies, where toxicities in animals did not prevent administration of these novel drugs to man. In this study, thrombocytopenias were the primary adverse human toxicities that were not detected by animal studies.

Hematology findings must be interpreted alongside other significant observations made within a study. In addition to detecting toxicity, preclinical animal studies also are useful for detecting exaggerated or unexpected pharmacodynamic effects at higher dosages than those proposed for new therapies. The number and relevance of the animal species used in toxicology studies remains a continual challenge (Zbinden, 1993), and the realization that not all compounds show a dose–response relationship gives an additional challenge to data interpretation (Calabrese and Baldwin, 2003).

No observable effects in animal toxicity studies does not imply absolute safety. In establishing safety margins, consideration must be given to the relationships between the no-observed-effect level (NOEL), no-observed-adverse-effect level (NOAEL), and the exposure levels expected in humans. Risks for human health must be considered in terms of exposure, toxicity, and balanced with the expected benefit. Because of the very nature of idiosyncratic adverse effects, their detection, where maybe only one in a million subjects is affected, remains a continual challenge. Unfortunately, there are several examples where preclinical testing has failed to ensure human safety. Hematology remains an important key science and gatekeeper in preclinical, clinical, and environmental safety.

REFERENCES

DATA STORAGE AND SECURITY

Food and Drug Administration. 2003. *Guidance for industry: Electronic records; electronic signatures. Scope and application.* 21 CFR 11. Center for Drug Evaluation and Research.

INTER- AND INTRAINDIVIDUAL VARIATION

Carakostas, M. C., and Banerjee, A. K. 1990. Interpreting rodent clinical laboratory data in safety assessment studies: Biological and analytical components of variation. *Fundam. Appl. Toxicol.* 15:744–53.

Gaylor, D. W., Suber, R. L., Wolff, G. L., and Crowell, J. A. 1987. Statistical variation of selected clinical pathological and biochemical measurements in rodents. *Proc. Soc. Exp. Biol. Med.* 185:361–67.

Weil, C. S. 1982. Statistical analysis and normality of selected hematologic and clinical chemistry measurements used in toxicologic studies. *Arch. Toxicol.* 5(Suppl.):237–53.

REFERENCE RANGE

IFCC. 1987. Approved recommendation (1987) on the theory of reference values. Part 5. Statistical treatment of collected reference values. Determination of reference limits. *Clin. Chim. Acta* 170:S13–32.

CRITICAL DIFFERENCES

Fraser, C. G., and Harris, E. G. 1989. Generation and application of data on biological variation in clinical chemistry. *CRC Crit. Rev. Clin. Lab. Sci.* 27:404–15.

Jensen, A. L., Iversen, L., and Petersen, T. K. 1998. Study on biological variability of hematological components in dogs. *Comp. Hematol. Int.* 8:202–4.

Jensen, A. L., Wenck, A., and Koch, J. 1994. Comparison of results of hematological and clinical chemical analyses of blood samples obtained from the cephalic and external jugular veins in dogs. *Res. Vet. Sci.* 56:24–29.

DESIGN AND SIZE OF ANIMAL STUDIES

Curtis, C. R., Salman, M. D., and Shott, S. 1990. Power and sample size. *JAVMA* 97:838–40.

Desu, M. M. 1990. *Sample size methodology.* London: Academic Press.

Festing, M. 1995. Are experiments in toxicological research "the right size"? In *Statistics in toxicology*, ed. B. J. T. Morgan. pp. 3–11, Oxford: Clarendon Press.

Festing, M. F. W., Overend, P., Das, R. G., Borja, M. C., and Berdoy, M. 2002. The design of animal experiments: Reducing the number of animals in research through better experimental design. *Laboratory animal handbook*, Vol. 14. London: Royal Society of Medicine Press.

Mann, M. D., Crouse, D. A., and Prentice, E. D. 1991. Appropriate animal numbers in biomedical research in light of animal welfare considerations. *Lab. Anim. Sci.* 41:6–14.

STATISTICAL APPROACHES

Dickens, A., and Robinson, J. 1996. Statistical approaches. In *Animal clinical chemistry: A primer for toxicologists*, ed. G. O. Evans. London: Taylor & Francis Ltd.

Festing, M. F. W., Hawkey, C. M., Hart, M. G., Turton, J. A., Gwynne, J., and Hicks, R. M. 1984. Principle components analysis of hematological data from F344 rats with bladder cancer fed N-(ethyl)-all-trans retinamide. *Food Chem. Toxicol.* 22:559–72.

Gad, S. C. 1998. *Statistics and experimental design for toxicologists.* 3rd ed. Boca Raton, FL: CRC Press.

Jones, P. K. 1988. R. A. Fisher and the 0.05 level of significance in medical studies. *J. Lab. Clin. Med.* 111:491–92.

Mills, J. L. 1993. Data torturing. *N. Engl. J. Med.* 329:1196–99.

Morgan, B. J. T., ed. 1996. *Statistics in toxicology.* Oxford: Clarendon Press.

Romano, P. E. 1988. The insignificance of a probability value of p<0.05 in the evaluation of medical scientific studies. *J. Lab. Clin. Med.* 111:501–3.

Waner, T. 1992. Current statistical approaches to clinical pathological data from toxicological studies. *Toxicol. Pathol.* 20:477–79.

Specificity and Sensitivity

Griner, P. F., Mayewski, R. J., Mushlin, A. I., and Greenland, P. 1981. Selection and interpretation of diagnostic tests and procedures. Principles and applications. *Ann. Intern. Med.* 94:553–600.

Predictivity

Calabrese, E. J., and Baldwin, L. A. 2003. Toxicology rethinks its central belief. *Nature* 421:691–92.

Garratini, S. 1985. Toxic effects of chemicals: Difficulties in extrapolating data from animals to man. *CRC Crit. Rev. Toxicol.* 16:1–29.

Heywood, R. 1981a. Target organ toxicity. *Toxicol. Lett.* 8:349–58.

Heywood, R. 1981b. Target organ toxicity. II. *Toxicol. Lett.* 18:83–88.

Olson, H., et al. 2000.Concordance of the toxicity of pharmaceuticals in humans and animals. *Reg. Toxicol. Pharmacol.* 32:56–67.

Schein, P. S., Davis, R. D., Carter, S., Newman, J., Schein, D. R., and Rall, D. P. 1970. The evaluation of anticancer drugs in dogs and monkeys for the prediction of qualitiative toxicities in man. *Clin. Pharmacol. Ther.* 11:3–40.

Schein, P., and Anderson, T. 1973. The efficacy of animal studies in predicting clinical toxicity of cancer therapeutic drugs. *Int. J. Clin. Pharmacol.* 8:228–38.

Zbinden, G. 1991. Predictive values of animal studies in toxicology. *Reg. Toxicol. Pharmacol.* 14:167–77.

Zbinden, G. 1993. The concept of multispecies testing in industrial toxicology. *Reg. Toxicol. Pharmacol.* 17:85–94.

Some Additional References for Statistical Tests

Armitage, P., and Berry, G. 1988. *Statistical methods in medical research.* 2nd ed. Oxford: Blackwell Scientific Publications.

Dunnett, C. W. 1955. A multiple comparison procedure for comparing several treatments with a control. *J. Am. Stat. Assoc.* 50:1096–121.

Dunnett, C. W. 1964. New table for multiple comparisons with a control. *Biometrics* 20:482–91.

Jonckheere, A. R. 1954. A distribution-free k-sample test against ordered alternatives. *Biometrika* 41:13–45.

Lee, P. N., and Lovell, D. P. 1999. Statistics for toxicology. In *General and applied toxicology,* ed. B. Ballantyne, T. Marrs, and P. Turner. 2nd ed. Basingstoke, England: Macmillan Reference Ltd.

Levene, H. 1960. *Contributions to probability and statistics.* Stanford, CA: Stanford University Press.

Shirley, E. 1977. A non-parametric equivalent of Williams' test for contrasting increasing dose levels of a treatment. *Biometrics* 33:386–89.

Snedcor, G. W., and Cochran, W. G. 1980. *Statistical methods.* 7th ed. Ames: Iowa State University Press.

Williams, D. A. 1972. The comparison of several dose levels with a zero dose control. *Biometrics* 28:519–31.

Appendix A: References for Laboratory Animal Hematology Data

For general information on laboratory animals, the Laboratory Animal Pocket Reference Series published by CRC Press provides information on biological data, husbandry, management, veterinary care, experimental methodology, and resources. The series includes *The Laboratory Rat, The Laboratory Non-human Primate, The Laboratory Rabbit, The Laboratory Hamster and Gerbil, The Laboratory Mouse, The Laboratory Guinea Pig*, and *The Laboratory Swine*.

REFERENCE VALUES

Throughout this book, attention has been drawn to the variables that affect reference range data. This author's perception is that automated hematology counters began to make an impact on measurements particularly for differential leukocyte counts in the more common laboratory animals species from about 1985. Additional data are found in the references provided at the end of each chapter.

ATLASES OF BLOOD CELL MORPHOLOGY

Most published atlases are of human morphology, but a few references that are helpful in animal hematology are listed here; however, there is no substitute for practical training within your own laboratory using the local techniques.

Feldman, B. F., Zinkl, J. G., and Jain, N. C., eds. 2000. *Schalm's veterinary hematology*. 5th ed. Philadelphia: Lippincott, Williams & Wilkins.

Hasegawa, A., and Furuhama, K. 1998. *Atlas of the hematology of the laboratory rat*. New York: Elsevier.

Hawkey, C. M., and Dennett, T. B. 1989. *A colour atlas of comparative veterinary haematology*. London: Wolfe Publishing.

Jones, T. C., Ward, J. M., Mohr, U., and Hunt, R. D., eds. 1990. *Hemopoietic system. Monographs on pathology of laboratory animals*. New York: Springer-Verlag.

Rowley, A. F., and Ratcliffe, N. A., eds. 1988. *Vertebrate blood cells*. Cambridge, England: Cambridge University Press.

Smith, C. A., Andrews, C. M., Collard, J. K., Hall, D. E., and Walker, A. K., eds. 1993. *Color atlas of comparative diagnostic & experimental hematology*. London: Wolfe Publishing, Mosby-Year Book Europe Ltd.

GENERAL (CONTAINS INFORMATION ON SEVERAL SPECIES)

Archer, R. K., and Jeffcott, L. B., eds. 1977. *Comparative clinical haematology.* Oxford: Blackwell Scientific Publications.

Bailly, Y., and Duprat, P. 1990. Normal blood cell values. Haemopoeitic system—Pathophysiology. In *Monographs on pathology of laboratory animals*, ed. T. C. Jones, J. M. Ward, U. Mohr, and R. D. Hunt, 27–33. Berlin: Springer-Verlag.

Ingerman, R. L. 1997. Vertebrate hemoglobins. In *Handbook of physiology: Comparative physiology*, ed. W. H. Dantzler, section 13, chap. 6. Vol. 1. New York: Oxford University Press.

Levine, B. S. 1995. Animal clinical pathology. In *CRC handbook of toxicology*, ed. M. J. Derelanko and M. A. Hollinger, 518–35. Boca Raton, FL: CRC Press.

Matsuzawa, T., Nomura, M., and Unno, T. 1993. Clinical pathology reference ranges of laboratory animals. *J. Vet. Med. Sci.* 55:351–62.

Mitruka, B. M., and Rawnsley, H. M. 1977. *Clinical biochemical and hematological reference values in normal experimental animals.* New York: Masson Publishing USA.

Payne, B. J., Lewis, H. B., Murchison, T. E., and Hart, E. A. 1976. Hematology of laboratory animals. In *Handbook of laboratory animal science*, ed. E. C. Melby and N. H. Altman, 383–461. Cleveland, OH: CRC Press.

Rowley, A. F., and Radcliffe, N. A. 1988. *Vertebrate blood cells.* Cambridge, England: Cambridge University Press.

Sanderson, J. H., and Phillips, C. E. 1981. *An atlas of laboratory animal haematology.* Oxford: Clarendon Press.

Wolford, S. T., Schroer, R. A., Gohs, F. X., Gallo, P. P., Brodeck, M., Falk, H. B., and Ruhren, R. 1986. Reference range data base for serum chemistry and hematology values in laboratory animals. *J. Toxicol. Env. Health* 8:161–88.

DOG, *CANIS FAMILIARIS*

Andersen, A. C., and Schalm, O. W. 1970. Hematology. In *The beagle as an experimental dog*, ed. A. C. Andersen and L. S. Good, 261–81. Ames: Iowa State University Press.

Catalfamo, J. L., and Dodds, W. J. 1988. Inherited and acquired thrombopathias. In *Veterinary clinics of North America. Small animal practice*, ed. B. F. Feldman, 185–93. Vol. 18. Orlando, FL: Elsevier.

Dougherty, J. H., and Rosenblatt, L. S. 1965. Changes in the hemogram of the beagle with age. *J. Gerontol.* 20:131–38.

Doxey, D. L. 1966. Some conditions associated with variations in circulating oestrogens— Blood picture alterations. *J. Small Anim. Prac.* 7:375–85.

Evans, G. O., Fagg, R., Flynn, R. M., and Smith, D. E. C. 1991. Haematological measurements in healthy beagle dogs using a Sysmex E-5000. *Sysmex J.* 14:119–20.

Evans, G. O., and Fagg, R. 1994. Reticulocyte counts in canine and rat blood made by flow cytometry. *J. Comp. Pathol.* 111:107–11.

Hall, D. E. 1972. *Blood coagulation and its disorders in the dog.* London: Balliere Tindall.

Michaelson, S. M., Scheer, K., and Gilt, S. 1966. The blood of the normal beagle. *JAVMA* 148:532–34.

Spurling, N. W. 1977. Haematology of the dog. In *Comparative clinical haematology*, ed. R. K. Archer and L. B. Jeffcott. 365–440. Oxford: Blackwell Scientific Publications.

Uchiyama, T., Tokoi, K., and Deki, T. 1985. Successive changes in the blood composition of the experimental normal beagle dogs accompanied with age. *Exp. Anim.* (Japan) 34:367–77.

FERRET, *MUSTELA PUTORIUS FURO*

Lee, E. J., Moore, W. E., Fryer, H. C., and Minocha, H. C. 1982. Haematological and serum chemistry profiles of ferrets (*Mustela putorius furo*). *Lab. Anim.* 16:133–37.

Marini, R. P., Jackson, L. R., Esteves, M. I., Andrutis, K. A., Goslant, C. M., and Fox, J. G. 1994. Effects of isoflurane on hematologic variables. *Am. J. Vet. Res.* 55:1479–83.

GUINEA PIG, *CAVIA PORCELLUS*

Kasparett, J., Messow, C., and Edel, J. 1988. Blood coagulation studies in guinea pigs (*Cavia porcellus*). *Lab. Anim.* 22:206–11.

Kitagaki, M., Yamaguchi, M., Nakamura, M., Sakurada, K., Suwa, T., and Sasa, H. 2005. Age-related changes in haematology and serum chemistry of Weiser-Maples guineapigs (*Cavia porcellus*). *Lab. Anim.* 39:321–30.

Lewis, J. H. 1992. Comparative haematology: Studies on guinea-pigs (*Cavia porcellus*). *Comp. Biochem. Physiol.* 102A:507–12.

Sisk, D. B. 1976. Physiology. In *The biology of the guinea pig*, ed. J. E. Wagner and P. J. Manner, 64–74. New York: Academic Press.

Waner, T., Avidar, Y., Peh, H. C., Zass, R., and Bogin, E. 1996. Haematology and clinical values of normal and euthymic adult hairless male Dunkin-Hartley guinea pigs (*Cavia porcellus*). *Vet. Clin. Pathol.* 25:61–64.

Yoshihara, K., Wanatabe, A., Inaba, T., Kuramoto, M., and Shiratori, K. 1995. A biological study of inbred Weiser-Maples guinea pigs—Urinalysis, hematological and blood chemical values and organ weights. *Exp. Anim.* 43:737–45.

HAMSTER,
MESOCRICETUS AURATUS (SYRIAN HAMSTER), CRICETULUS GRISEUS (CHINESE HAMSTER)

Dontenwill, W., Chevalier, H.-J., Harke, H.-P., Lafrenz, U., Reckzeh, G., and Leuschner, F. 1974. Biochemical and haematological investigations in Syrian golden hamsters after cigarette smoke inhalation. *Lab. Anim.* 8:217–35.

Tomson, F. N., and Wardrop, K. J. 1987. Clinical chemistry and hematology. In *Laboratory hamsters*, ed. G. L. van Hoosier and C. W. McPherson, 43–59. Orlando, FL: Academic Press.

MARMOSET, *CALLITHRIX JACCHUS*

Anderson, E. T., Lewis, J. P., Passovoy, M., and Trobaugh, F. E. 1967. Marmosets as laboratory animals. II. The hematology of laboratory kept marmosets. *Lab. Anim. Care* 17:30–40.

Evans, G. O., and Smith, D. E. C. 1986. Platelet measurements in the common marmoset *Callithrix jacchus*. *J. Comp. Pathol.* 96:343–47.

Hawkey, C. M., Hart, M. G., and Jones, D. M. 1982. Clinical hematology of the common marmoset *Callithrix jacchus*. *Am. J. Primatol.* 3:179–99.

Yarbrough, L. W., Tollett, J. L., Montrey, R. D., and Beattie, R. J. 1984. Serum biochemical hematological and body measurement data for common marmosets (*Callithrix jacchus*). *Lab. Anim. Sci.* 34:276–80.

MOUSE, *MUS*

Bannerman, R. M. 1983. Hematology. In *The mouse in biomedical research* Vol. III, eds. Foster, H., Small J. D., and Fox J. G. 291–312. New York: Academic Press.

Cotchin, E., and Roe, F. J. C. 1967. *Pathology of laboratory rats and mice.* Oxford: Blackwell Scientific Publications.

Duarte, A. P. T., Ramwell, P., and Myers, A. 1986. Sex differences in mouse platelet aggregation. *Thromb. Res.* 43:33–39.

Frith, C. H., Suber, R. L., and Umholtz, R. 1980. Hematologic and clinical chemistry findings in control BALB/c and C57BL/6 mice. *Lab. Anim. Sci.* 30:835–40.

Kajioka, E. H., Andres, M. L., Nelson, G. A., and Gridley, D. S. 2000. Immunologic variables in male and female C57BL/6 mice from two sources. *Comp. Med.* 50:288–91.

Russell, E. S., Neufeld, E. F., and Higgins, C. T. 1951. Comparison of normal blood picture of young adults from 18 inbred strains of mice. *Proc. Soc. Exp. Biol. Med.* 78:761–66.

Schneck, K., Washington, M., Holder, D., Lodge, K., and Motzel, S. 2000. Hematological and serum reference values in non-transgenic FVB mice. *Comp. Med.* 50:32–35.

Tsakiris, D. A., Scudder, L., Hodivala-Dilke, K., Hynes, R. O., and Coller, B. S. 1999. Hemostasis in the mouse (*Mus musculus*): A review. *Thromb. Haemost.* 81:177–88.

Nonhuman Primate,

Macaca fascicularis (Cynomolgus) and Macaca artoides (Stump-Tailed Monkeys)

Buchl, S. J., and Howard, B. 1997. Hematologic and serum biochemical and electrolyte values in clinically normal domestically bred Rhesus monkeys (*Macaca mulatta*) according to age sex and gravidity. *Lab. Anim. Sci.* 47:528–33.

de Neef, K. J., Nieuwenhuijsen, K., Lammers, A. J. J. C., Degen, A. J. M., and Verbon, F. 1987. Blood variables in adult stumptail macques (*Macaca artoides*) living in a captive group: Annual variability. *J. Med. Primatol.* 16:237–47.

Huser, H.-J. 1970. *Atlas of comparative primate hematology.* New York: Academic Press.

Matsumoto, K., Akagi, H., Ochiai, T., Hagino, K., Sekita, K., Kawasaki, Y., Matin, M. A., and Furuya, T. 1980. Comparative blood values of *Macaca mulatta* and *Macaca fascicularis.* *Exp. Anim.* 29:335–40.

Matsuzawa, T., and Nagia, Y. 1994. Comparative haematological and plasma chemistry values in purpose-bred Squirrel, Cynomolgus and Rhesus monkeys. *Comp. Haematol. Int.* 4:43–48.

Verlangieri, A. J., DePriest, J. C., and Kapeghian, J. C. 1985. Normal serum biochemical, haematological and EKG parameters in anethetized adult male *Macaca fascicularis* and *Macaca artoides.* *Lab. Anim. Sci.* 35:63–66.

Micropigs/Minipigs

Ellegaard, L., Damm-Jorgensen, S., Klastrup, S., Kornerup-Hansen, A., and Svendsen, O. 1995. Hematological and clinical chemistry values in 3 and 6 months old Göttingen minipigs. *Scand. J. Lab. Clin. Anim. Sci.* 22:239–48.

Petroianu, G., Maleck, W., Altmannsberger, S., Jatzko, A., and Rüfer, R. 1997. Blood coagulation, platelets and haematocrit in male, female and pregnant Göttingen minipigs. *Scand. J. Lab. Anim. Sci.* 24:31–41.

Radin, M. J., Weiser, M. G., and Fettman, M. J. 1986. Hematologic and serum biochemical values for Yucatan minature swine. *Lab. Anim. Sci.* 36:425–27.

Rispat, G., Slaoui, M., Weber, D., Salemink, P., Berthoux, C., and Shrivastava, R. 1993. Haematological and plasma biochemical values for healthy Yucatan micropigs. *Lab. Anim.* 27:368–73.

MONGOLIAN GERBIL, *MERIONES UNGUICULATUS*

Dillon, W. G., and Glomski, C. A. 1975. The Mongolian gerbil: Qualitative and quantitative aspects of the cellular blood picture. *Lab. Anim.* 9:283–87.

Mays, A. 1969. Baseline hematological and blood biochemical parameters of the Mongolian gerbil (*Meriones unguiculatus*). *Lab. Anim. Care* 19:838–42.

RABBIT, *ORYCTOLAGUS CUNICULUS*

Bortolotti, A., Castelli, D., and Bonati, M. 1989. Hematology and serum chemistry values of adult, pregnant and newborn New Zealand rabbits (*Oryctolagus cuniculus*). *Lab. Anim. Sci.* 39:437–39.

Dubiski, S., ed. 1987. *The rabbit in contemporary immunological research*. London: Longman Group UK Ltd.

Hewitt, C. D., Innes, D. J., Savory, J., and Wills, M. R. 1989. Normal biochemical and haematological values in New Zealand white rabbits. *Clin. Chem.* 35:1777–79.

Kabata, J., Gratwohl, A., Tichelli, A., John, L., and Speck, B. 1991. Hematologic values of New Zealand White rabbits determined by automated flow cytometry. *Lab. Anim. Sci.* 41:613–19.

Kriesten, K., Murawski, U., and Schmidtmann, W. 1987. Haematological values during normal reproduction of the maternal and the fetal rabbit. *Comp. Biochem. Physiol.* 87A:479–85.

McLaughlin, R. M., and Fish, R. E. 1994. Clinical biochemistry and hematology. In *The biology of the laboratory rabbit*, ed. P. J. Manning, D. H. Ringler, and C. E. Newcomer, 111–24. 2nd ed. San Diego: Academic Press.

Wills, J. E., Rowlands, M. A., North, D. C., and Evans, G. O. 1993. Effects of serial cardiac puncture and blood collection procedures in the rabbit. *Anim. Tech.* 44:39–52.

RAT, *RATTUS NORVEGICUS*

Archer, R. K., Festing, M. F. W., and Riley, J. 1982. Haematology of conventionally-maintained Lac:P outbred Wistar rats during the 1st year of life. *Lab. Anim.* 16:198–200.

Cotchin, E., and Roe, F. J. C. 1967. *Pathology of laboratory rats and mice*. Oxford: Blackwell Scientific Publications.

Edwards, C. J., and Fuller, J. 1992. Notes on age related changes in differential leucocyte counts of the Charles River Outbred albino SD rat and CD1 mouse. *Comp. Haematol. Int.* 2:58–64.

Evans, G. O., Fagg, R., and Smith, D. E. C. 1993. Manual absolute reticulocyte counts in healthy Wistar rats. *Comp. Haematol. Int.* 3:116–18.

Evans, G. O., and Fagg, R. 1994. Reticulocyte counts in canine and rat blood made by flow cytometry. *J. Comp. Pathol.* 111:107–11.

Godwin, K. O., Fraser, F. J., and Ibbotson, R. N. 1964. Haematological observations on healthy (SPF) rats. *Br. J. Exp. Pathol.* 45:514–24.

Hackbarth, H., Burow, K., and Schimansky, G. 1983. Strain differences in inbred rats: Influence of strain and diet on haematological traits. *Lab. Anim.* 17:7–12.

Kojima, S., Haruta, J., Enomoto, A., Fujisawa, H., Harada, T., and Maita, K. 1999. Age related hematological changes in normal F344 rats: During the neonatal period. *Exp. Anim.* 48:153–59.

Kozma, C. K., Weisbroth, S. H., Stratman, S. L., and Conejeros, M. 1969. Normal biological values for Long-Evans rats. *Lab. Anim. Care* 19:747–55.

Leonard, R., and Ruben, Z. 1986. Hematology reference values for peripheral blood of laboratory rats. *Lab. Anim. Sci.* 36:277–81.

Lewi, P. J., and Marsboom, R. P. 1981. *Toxicology reference data—Wistar rat.* Amsterdam, Holland: Elsevier/North-Holland Biomedical Press.

Lillie, L., Temple, N. J., and Florence, L. Z. 1996. Reference values for young normal Sprague-Dawley rats: Weight gain, hematology and clinical chemistry. *Hum. Exp. Toxicol.* 15:612–16.

Lovell, D. P., Archer, R. K., Riley, J., and Morgan, R. K. 1981. Variation in haematological parameters among inbred strains of rat. *Lab. Anim.* 15:243–49.

Oyekan, A. O., and Botting, J. H. 1991. Relationship between gender difference in intravascular aggregation of platelets and the fibrinolytic pathway in the rat. *Arch. Int. Pharmacodyn.* 313:176–92.

Papworth, T. A., and Clubb, S. K. 1995. Clinical pathology in the neonatal rat. *Comp. Haematol. Int.* 5:237–50.

Pettersen, J. C., Morrissey, R. L., Saunders, D. R., Pavkov, K. L., Luempert, L. G., Turnier, J. C., Matheson, D. W., and Schwartz, D. R. 1996. A 2-year comparison study of Crl:CD BR and Hsd:Sprague-Dawley SD rats. *Fundam. Appl. Toxicol.* 33:196–211.

Reich, C., and Dunning, W. F. 1942. Studies on the morphology of the peripheral blood of rats. *Cancer Res.* 3:248–57.

Ringler, D. H., and Dabich, L. 1979. Hematology and clinical biochemistry. In *The laboratory rat,* ed. H. J. Baker, J. R. Lindsey, and S. H. Weisbroth, 105–21. Vol. 1. Orlando, FL: Academic Press.

Smith, D., and Bronson, R. 1992. Clinical chemistry and hematology profiles of the aging rat. *Lab. Anim.* 21:32–46.

Stromberg, P. C. 1992. Changes in the hematologic system. In *Pathobiology of the aging rat,* ed. U. Mohr, D. L. Dungworth, and C. C. Capen, 15–24. Washington, DC: ILSI Press.

Turton, J. W., Hawkey, C. M., Hart, M. G., Gwynne, J., and Hicks, R. M. 1989. Age-related changes in the haematology of female F344 rats. *Lab. Anim.* 23:295–301.

Wolford, S. T., Schroer, R. A., Gallo, P. P., Gohs, F. X., Brodeck, M., Falk, H. B., and Ruhren, R. 1987. Age-related changes in serum chemistry and hematology values in normal Sprague-Dawley rats. *Fundam. Appl. Toxicol.* 8:80–88.

Appendix B:
SI Units and Conversions

The International System of Units (Systeme International [SI]) was adopted in 1960 by the General Conference of Weights and Measures as a coherent system based on seven basic units: meter, kilogram, second, ampere, Kelvin, candela, and mole. Some of the common hematological units are provided here with factors to convert from the traditional non-SI units to SI unitage. In human medicine, the system has not been adopted universally, and even for hemoglobin we still encounter values expressed as g/dl, g/l, mmol/l, and %, with various options for reporting units provided with analyzers. There is added confusion in the conversion of hemoglobin to molar concentrations, where some laboratories assume the molar mass is 16,000 (multiply g/dl by 0.62 to obtain mmol/l) and other laboratories use a molar mass of 64,000 (multiply g/dl by 0.155 to obtain mmol/l).

REFERENCES

Baron, D. N., Broughton, P. M. G., Cohen, M., Lansley, T. S., Lewis, S. M., and Shinton, N. K. 1974. The use of SI units in reporting results obtained in hospital laboratories. *J. Clin. Pathol.* 27:590–97.

Blomback, M., Dykbaer R., Jorgensen II., Olesen H., and Thorsen, S. 1995. Properties and units in the clinical laboratory science. V. Properties and units in thrombosis and haemostasis. ISTH-IUPAC-IFCC recommendations. *Clin. Chim. Acta* 245:S5–S22.

Editorial. 1986. Now read this: The SI units are here. *J. Am. Med. Assoc.* 255:2329–39.

Laposata, M. 1992. *SI unit conversion guide.* Waltham: Massachusetts Medical Society (NEJM).

Olesen, H. 1996. Properties and units in the clinical laboratotory science. 1. Syntax and semantic rules. IUPAC-IFCC recommendations 1995. *Clin. Chim. Acta* 245:S5–22.

fl or fL	Femtoliter (10^{-15}/l)	fg	Femtogram	fmol	Femtomole
pl or pL	Picoliter (10^{-12}/l)	pg	Picogram	pmol	Picomole
nl or nL	Nanoliter (10^{-9}/l)	ng	Nanogram	pmol	Picomole
μl or μL	Microliter (10^{-6}/l)	μg	Microgram	μmol	Micromole
ml	Milliliter (10^{-3}/l)	mg	Milligram	mmol	Millimole
dl or dL	Deciliter (10^{-1}/l)				

M	Molar concentration
N	Normal concentration
Mol.kg	Molal concentration
kD or KDa	Kilodaltons

Test	Abbreviation	SI Units	Traditional Units	Multiply to Convert to SI Units
Hematocrit/packed cell volume	Hct (or PCV)	Ratio	%	0.01
Hemoglobin	Hb or Hgb	g/dl (g/l)	g/dl	1 (or 10)
Mean corpuscular hemoglobin	MCH	pg	pg	1.0
Mean corpuscular hemoglobin concentration	MCHC	g/dl	%	1.0
Mean corpuscular volume	MCV	fl	μm^3	1.0
Erythrocyte counts	RBC	$10^{12}/l$	$10^6/\mu l$ or millions/μl or cu.mm	10^6
Leukocyte counts	WBC	$10^9/l$	Thousands/μl or cu.mm	10^6
Platelet counts	PLT or PLAT	$10^9/l$	Thousands/μl or cu.mm	10^6
Reticulocyte counts	RETC or Retic	%	%	1.0
		$10^9/l$	Thousands/μl or cu.mm	10^6

Test	SI Units	Traditional Units	Multiply to Convert to SI Units
Cobalamin (vitamin B12)	pmol/l	*pg/ml*	0.738
Fibrinogen	Mg/l	*mg/dl*	0.01
Iron	μmol/l	*$\mu g/dl$*	0.179
Folic acid	nmol/l	*ng/ml*	2.27

Appendix C: Expectable Ranges

Investigators coming into the field of laboratory animal hematology sometimes ask: "What values might I expect to encounter in the healthy animals?" From the information provided in Chapters 9 and 10, the reader will understand that these ranges can be affected by many preanalytical and analytical variables in the measurements, and it must be emphasized that the local laboratory should be asked to comment on its own ranges and experience.

The following tables are provided to indicate the ranges of values that might be obtained for several species of laboratory animals. These ranges are based on published and unpublished data sources, and they should be treated as guides. These "expectable ranges" and the references provided in Appendix A can be used for comparisons when establishing your locally determined ranges. Additional blank table columns have been included to allow readers to insert their local ranges for reference.

Ranges for Dogs (Beagles) Aged 8 to 12 Months

Measurement	Range	
Erythrocytic Data		
Erythrocyte count (RBC) (10^{12}/l)	5 to 8	
Hemoglobin (g/dl)	12 to 18	
Hematocrit (ratio)	0.36 to 0.52	
Mean cell volume (MCV) (fl)	58 to 73	
Mean cell hemoglobin (MCH) (pg)	20 to 25	
Mean cell hemoglobin concentration (MCHC) (g/dl)	32 to 36	
Reticulocytes (% RBC)	0 to 3	
Reticulocytes (10^9/l)	10 to 150	
Leukocytic Data		
Total leukocyte count (WBC) (10^9/l)	5 to 20	
Lymphocytes (10^9/l)	1 to 7	
Neutrophils (10^9/l)	3 to 12	
Monocytes (10^9/l)	0.1 to 2	
Eosinophils (10^9/l)	0.1 to 2	
Basophils (10^9/l)	Rare	
Lymphocytes (%)	15 to 45	
Neutrophils (%)	45 to 80	
Monocytes (%)	0.1 to 8	
Eosinophils (%)	0.1 to 8	
Basophils (%)	Rare	
Platelet Data		
Platelet count (10^9/l)	200 to 500	

Ranges for Ferrets

Measurement	Range	
Erythrocytic Data		
Erythrocyte count (RBC) (10^{12}/l)	7 to 11	
Hemoglobin (g/dl)	14 to 18	
Hematocrit (ratio)	0.40 to 0.55	
Mean cell volume (MCV) (fl)	48 to 55	
Mean cell hemoglobin (MCH) (pg)	15 to 20	
Mean cell hemoglobin concentration (MCHC) (g/dl)	30 to 36	
Reticulocytes (% RBC)	—	
Reticulocytes (10^9/l)	—	
Leukocytic Data		
Total leukocyte count (WBC)(10^9/l)	4 to 16	
Lymphocytes (10^9/l)	2 to 7	
Neutrophils (10^9/l)	2 to 11	
Monocytes (10^9/l)	0 to 0.8	
Eosinophils (10^9/l)	0 to 0.8	
Basophils (10^9/l)	0 to 0.2	
Lymphocytes (%)	25 to 70	
Neutrophils (%)	10 to 70	
Monocytes (%)	1 to 5	
Eosinophils (%)	0.1 to 4	
Basophils (%)	0 to 1	
Platelet Data		
Platelet count (10^9/l)	300 to 1200	

Ranges for Guinea Pigs

Measurement	Range	
Erythrocytic Data		
Erythrocyte count (RBC) (10^{12}/l)	3 to 6.5	
Hemoglobin (g/dl)	11 to 16	
Hematocrit (ratio)	0.35 to 0.50	
Mean cell volume (MCV) (fl)	78 to 88	
Mean cell hemoglobin (MCH) (pg)	25 to 30	
Mean cell hemoglobin concentration (MCHC) (g/dl)	28 to 34	
Reticulocytes (% RBC)	1 to 5	
Reticulocytes (10^9/l)	50 to 250	
Leukocytic Data		
Total leukocyte count (WBC) (10^9/l)	3 to 17	
Lymphocytes (10^9/l)	1 to 15	
Neutrophils (10^9/l)	1 to 15	
Monocytes (10^9/l)	0.1 to 1	
Eosinophils (10^9/l)	0.1 to 1	
Basophils (10^9/l)	Rare	
Lymphocytes (%)	20 to 70	
Neutrophils (%)	40 to 75	
Monocytes (%)	0 to 7	
Eosinophils (%)	0 to 5	
Basophils (%)	Rare	
Platelet Data		
Platelet count (10^9/l)	250 to 1,000	

Ranges for Hamster

Measurement	Range	
Erythrocytic Data		
Erythrocyte count (RBC) (10^{12}/l)	4 to 10	
Hemoglobin (g/dl)	12 to 17	
Hematocrit (ratio)	0.40 to 0.55	
Mean cell volume (MCV) (fl)	55 to 70	
Mean cell hemoglobin (MCH) (pg)	17 to 24	
Mean cell hemoglobin concentration (MCHC) (g/dL)	30 to 35	
Reticulocytes (% RBC)	1 to 5	
Reticulocytes (10^9/l)	50 to 250	
Leukocytic Data		
Total leukocyte count (WBC) (10^9/l)	4 to 12	
Lymphocytes (10^9/l)	2 to 10	
Neutrophils (10^9/l)	0.1 to 4	
Monocytes (10^9/l)	0.1 to 2	
Eosinophils (10^9/l)	0.1 to 1	
Basophils (10^9/l)	Rare	
Lymphocytes (%)	50 to 90	
Neutrophils (%)	15 to 50	
Monocytes (%)	0 to 4	
Eosinophils (%)	0 to 2	
Basophils (%)	Rare	
Platelet Data		
Platelet count (10^9/l)	400 to 1,100	

Ranges for Marmosets

Measurement	Range	
Erythrocytic Data		
Erythrocyte count (RBC) (10^{12}/l)	5 to 9	
Hemoglobin (g/dl)	11 to 18	
Hematocrit (ratio)	0.40 to 0.57	
Mean cell volume (MCV) (fl)	63 to 77	
Mean cell hemoglobin (MCH) (pg)	20 to 25	
Mean cell hemoglobin concentration (MCHC) (g/dl)	28 to 35	
Reticulocytes (% RBC)	0 to 4	
Reticulocytes (10^9/l)	50 to 400	
Leukocytic Data		
Total leukocyte count (WBC) (10^9/l)	4 to 15	
Lymphocytes (10^9/l)	1 to 8	
Neutrophils (10^9/l)	1 to 8	
Monocytes (10^9/l)	0 to 1	
Eosinophils (10^9/l)	0 to 1	
Basophils (10^9/l)	0 to 0.5	
Lymphocytes (%)	30 to 80	
Neutrophils (%)	20 to 65	
Monocytes (%)	0 to 5	
Eosinophils (%)	0 to 2	
Basophils (%)	0 to 1	
Platelet Data		
Platelet count (10^9/l)	250 to 800	

Ranges for Mice Aged 6 to 10 Weeks

Measurement	Range	
Erythrocytic Data		
Erythrocyte count (RBC) (10^{12}/l)	7 to 11	
Hemoglobin (g/dl)	12 to 17	
Hematocrit (ratio)	0.38 to 0.50	
Mean cell volume (MCV) (fl)	44 to 52	
Mean cell hemoglobin (MCH) (pg)	15 to 19	
Mean cell hemoglobin concentration (MCHC) (g/dl)	32 to 36	
Reticulocytes (% RBC)	1 to 6	
Reticulocytes (10^9/l)	80 to 280	
Leukocytic Data		
Total leukocyte count (WBC) (10^9/l)	4 to 12	
Lymphocytes (10^9/l)	2.5 to 11	
Neutrophils (10^9/l)	0.3 to 2.5	
Monocytes (10^9/l)	0.1 to 1	
Eosinophils (10^9/l)	0.1 to 1	
Basophils (10^9/l)	Rare	
Lymphocytes (%)	60 to 90	
Neutrophils (%)	10 to 40	
Monocytes (%)	0 to 6	
Eosinophils (%)	0 to 3	
Basophils (%)	Rare	
Platelet Data		
Platelet count (10^9/l)	800 to 1,600	

Ranges for Monkeys (Cynomolgus)

Measurement	Range	
Erythrocytic Data		
Erythrocyte count (RBC) (10^{12}/l)	4 to 8	
Hemoglobin (g/dl)	11 to 14	
Hematocrit (ratio)	0.34 to 0.46	
Mean cell volume (MCV) (fl)	55 to 70	
Mean cell hemoglobin (MCH) (pg)	18 to 25	
Mean cell hemoglobin concentration (MCHC) (g/dl)	25 to 36	
Reticulocytes (% RBC)	1 to 2	
Reticulocytes (10^9/l)	50 to 200	
Leukocytic Data		
Total leukocyte count (WBC) (10^9/l)	3 to 18	
Lymphocytes (10^9/l)	2 to 10	
Neutrophils (10^9/l)	1 to 13	
Monocytes (10^9/l)	0.1 to 1	
Eosinophils (10^9/l)	0.1 to 2	
Basophils (10^9/l)	0.1 to 0.4	
Lymphocytes (%)	20 to 85	
Neutrophils (%)	15 to 80	
Monocytes (%)	0 to 5	
Eosinophils (%)	0 to 7	
Basophils (%)	0 to 5	
Platelet Data		
Platelet count (10^9/l)	200 to 700	

Ranges for Micropigs

Measurement	Range	
Erythrocytic Data		
Erythrocyte count (RBC) (10^{12}/l)	5 to 9	
Hemoglobin (g/dl)	11 to 16	
Hematocrit (ratio)	0.36 to 0.55	
Mean cell volume (MCV) (fl)	54 to 70	
Mean cell hemoglobin (MCH) (pg)	19 to 25	
Mean cell hemoglobin concentration (MCHC) (g/dl)	30 to 36	
Reticulocytes (% RBC)	1 to 6	
Reticulocytes (10^9/l)	—	
Leukocytic Data		
Total leukocyte count (WBC) (10^9/l)	7 to 20	
Lymphocytes (10^9/l)	—	
Neutrophils (10^9/l)	—	
Monocytes (10^9/l)	—	
Eosinophils (10^9/l)	—	
Basophils (10^9/l)	—	
Lymphocytes (%)	15 to 75	
Neutrophils (%)	20 to 70	
Monocytes (%)	2 to 12	
Eosinophils (%)	0.1 to 7	
Basophils (%)	0.1 to 3	
Platelet Data		
Platelet count (10^9/l)	200 to 800	

Ranges for Rabbits (New Zealand White)

Measurement	Range	Notes
Erythrocytic Data		
Erythrocyte count (RBC) (10^{12}/l)	4.5 to 7.5	
Hemoglobin (g/dl)	11 to 15	
Hematocrit (ratio)	0.32 to 0.50	
Mean cell volume (MCV) (fl)	62 to 72	
Mean cell hemoglobin (MCH) (pg)	19 to 24	
Mean cell hemoglobin concentration (MCHC) (g/dl)	31 to 35	
Reticulocytes (% RBC)	1 to 5	
Reticulocytes (10^9/l)	80 to 250	
Leukocytic Data		
Total leukocyte count (WBC) (10^9/l)	5 to 16	
Lymphocytes (10^9/l)	2 to 12	
Neutrophils (10^9/l)	2 to 8	
Monocytes (10^9/l)	0 to 1	
Eosinophils (10^9/l)	0 to 1	
Basophils (10^9/l)	0 to 1	
Lymphocytes (%)	20 to 90	
Neutrophils (%)	15 to 60	
Monocytes (%)	0 to 10	
Eosinophils (%)	0 to 5	
Basophils (%)	0 to 6	
Platelet Data		
Platelet count (10^9/l)	200 to 700	

Ranges for Rats (Wistar) Aged 7 to 14 Weeks

Measurement	Range	
Erythrocytic Data		
Erythrocyte count (RBC) (10^{12}/l)	6 to 9	
Hemoglobin (g/dl)	11 to 17	
Hematocrit (ratio)	0.38 to 0.50	
Mean cell volume (MCV) (fl)	50 to 62	
Mean cell hemoglobin (MCH) (pg)	17 to 22	
Mean cell hemoglobin concentration (MCHC) (g/dl)	31 to 36	
Reticulocytes (% RBC)	1 to 4	
Reticulocytes (10^9/l)	50 to 350	
Leukocytic Data		
Total leukocyte count (WBC) (10^9/l)	4 to 17	
Lymphocytes (10^9/l)	3 to 15	
Neutrophils (10^9/l)	0 to 4	
Monocytes (10^9/l)	0 to 2	
Eosinophils (10^9/l)	0 to 0.5	
Basophils (10^9/l)	Rare	
Lymphocytes (%)	65 to 90	
Neutrophils (%)	5 to 30	
Monocytes (%)	1 to 6	
Eosinophils (%)	0 to 3	
Basophils (%)	Rare	
Platelet Data		
Platelet count (10^9/l)	800 to 1,400	

Appendix D: General Abbreviations

This list includes some of the common acronyms used in toxicology and drug development, but every organization invents its own alphabet soup of acronyms, so beware!

AACC	American Association of Clinical Chemistry
ABPI	Association of British Pharmaceutical Industry
ADME	Absorption, distribution, metabolism, and excretion
ADR	Adverse drug reaction
ANOVA	Analysis of variance
ANSI	American National Standards Institute
AUC	Area under the curve; plasma concentration–time curve
BP or Bp	Blood pressure
CAS	Chemical Abstract Service
CDC	Centers for Disease Control
CFR	Code of Federal Regulations
CFSAN	Center for Food Safety and Applied Nutrition
CHIPS	Chemical Hazards Information and Packaging Supply
CL	Confidence limits
CNS	Central nervous system
CPMP	Committee for Proprietary Medicinal Products (EEC)
CRO	Contract research organization
CTC	Clinical trials certificate
CTM	Clinical trials material
CV	Coefficient of variation
CVS	Cardiovascular system
DACC	Division of Animal Clinical Chemistry, AACC
ECCP	European Comparative Clinical Pathology
ECG	Electrocardiogram
EDI	Estimated daily intake
EEC	European Economic Community
ELISA	Enzyme-linked immunoabsorbent assay
EMA	European Medicines Agency
EMEA	European Medicines Evaluation Agency
EPA	Environmental Protection Agency
EQAS	External quality assessment scheme
EU	European Union
FAO	Food and Agricultural Organization of the United Nations

FASEB	Federation of American Societies for Experimental Biology
FDA	Food and Drug Administration
FD&C	Food, Drug, and Cosmetic Act
FIFRA	Federal Insecticide, Fungicide, and Rodenticide Act
FOB	Functional observational battery
GCP	Good clinical practice
GI	Gastrointestinal
GIT	Gastrointestinal tract
GLP	Good laboratory practice
GMP	Good manufacturing practice
HPLC	High-performance liquid chromatography
HSE	Health and Safety Executive
IARC	International Agency for Research on Cancer
IC	Incapacitating concentration
IC_{50}	Concentration causing (or calculated to cause) 50% incapacitation of the population studied, e.g., cells
ICH	International Committee on Harmonization of Technical Requirements of Pharmaceuticals for Human Use
ID_{50}	Dose causing 50% inhibition in the population studied
IM	Intramuscular
IND	Investigational new drug
IP	Intraperitoneal
IPCS	International Program on Chemical Safety, WHO
IRIS	Integrated Risk Information System
ISO	International Standards Organization
IQ	Installation qualification
IV	Intravenous
JPMA	Japan Pharmaceutical Manufacturers Association
LAA	Laboratory animal allergy
LOAEL	Lowest-observed-adverse-effect level
LOD	Limit of detection
LOEL	Lowest-observed-effect level
LOQ	Limit of quantification
MAA	Marketing Authorization Authority
MCA	Medicines Control Agency (now MHRA)
MDD	Maximum daily dose
MHRA	Medicines and Healthcare Products Regulatory Agency
MHW	Ministry of Health and Welfare
MOA	Mode of action
MRC	Medical Research Council
MSDS	Material safety data sheet
MTD	Maximum tolerated dose
MW	Molecular weight
NCE	New chemical entity
NICE	National Institute for Clinical Excellence
NIEHS	National Institute of Environmental Health and Safety

NIOSH	National Institute of Occupational Safety and Health
NOAEL	No-observed-adverse-effect level
NOEL	No-observed-effect level
NONS	Notification of new substances
NTP	National Toxicology Program
OECD	Organization for Economic Cooperation and Development
OEL	Occupational exposure limit
OSHA	Occupational Safety and Health Administration
OTC	Over-the-counter sale
PC	Percutaneous
PCD	Programmed cell death
PEL	Permissible or permitted exposure level
PMA	Pharmaceutical Manufacturers Association
PMN	Premanufacturing notification
PMS	Postmarketing surveillance
PO	Peroral
QA	Quality assurance
QC	Quality control
QSAR	Quantitative structure-activity relationship
RTECS	Registry of Toxic Effects of Chemical Substances
SAR	Structure-activity relationship
SC	Subcutaneous
S.D.	Standard deviation
SE	Standard error
SEM	Standard error of the mean
SOP	Standard operating procedure
TDI	Total dietary intake
TLV	Threshold limit value
TSCA	Toxic Substances Control Act
USDA	U.S. Department of Agriculture
US EPA	U. S. Environmental Protection Agency
USP	U. S. Pharmacopica
VMD	Veterinary Medicine Directorate
WHO	World Health Organization

Appendix E: Some Common Biochemical Abbreviations

ADP	Adenosine diphosphate
ATP	Adenosine triphosphate
DNA	Deoxyribose nucleic acid
DPG	Diphosphoglycerate or diphosphoglyceric acid
EDTA	Ethylenediaminetetraacetic acid
EM pathway	Emden–Meyerhof pathway
EPO	Erythropoietin
ESF	Erythropoietin stimulating factor
F6P	Fructose-6-phosphate
G6PD	Glucose-6-phosphate dehydrogenase
GSH	Reduced glutathione
GSSG	Oxidized glutathione
IG or Ig	Immunoglobulin
IL	Interleukin
LDH	Lactic dehydrogenase
LPS	Lipopolysaccharide
NAD	Nicotinamide adenine dinucleotide
NADH	Reduced nicotinamide adenine dinucleotide
NADPH	Reduced nicotinamide adenine dinucleotide phosphate

Appendix F

All of these drugs, which are cited in the literature as causing an effect on blood cells, are used in several therapeutic areas—for example, the central nervous system, thyroid disorders, cancer, cardiovascular, antimicrobial, antiviral therapies, etc. It must be recognized that many of the published citations refer to a very small number of individual cases where adverse reactions have been reported. What the literature lacks is sufficient information about these drugs in the preclinical stages, which allows some correlation of the clinical and preclinical findings, and information on novel compounds, which allow some structural relationships to be investigated.

A continuing major information source for interactions of compounds on clinical laboratory tests is *Young's Effects Online* at www.fxol.org.

COMPOUNDS CAUSING REDUCTIONS AND DESTRUCTION OF ERYTHROCYTES

Acetahexamide
Acetaminophen
Acetanilid
Acetylphenylhydrazine
Acetylsalicylic acid
Acyclovir
Aminobenzoic acid
Aminopyrine
Aminosalicylic acid
Amidopurine
Amodiaquine
Amphetamine
Ampicillin
Amyl nitrite
Aniline
Antimony salts
Antineoplastic agents
Antipyrine
Aspirin
Azathioprine
Barbiturates
Benzene
Benzocaine

Bismuth salts
Busulfan
Carbamazepine
Carbenicillin
Cephalosporins (some)
Chlorambucil
Chloraphenicol
Chlordane
Chlorophenothane
Chloroquine
Chlorpromazine
Chlorpropamide
Chlortetracycline
Chlorthalidone
Colchicine
Cyclophosphamide
Cyclosporin
Cytarabine
Dactinomycin
Dapsone
Dicumarol
Digitalis
Diiodohydroxyquin

Dimercaprol
Dimethidene
Dinitrophenol
Diphenylhydramine
Dipyrone
Estrogens
Ethanol
Ethosuximide
Ethotoin
Fava beans
Fenoprofen
Floxuridine
Flucytosine
Furadaltone
Furazolidone
Furosemide
Gentamicin
Glucosulfone
Glutethimide
Gold salts
Haloperidol
Hexachlorobenzene
Hydrazaline

Hydrochlorothiazide
Hydroflumethiazide
Hydroxychloroquine
Hydroxyurea
Ibuprofen
Indomethacin
Iproniazid
Isocarboxazid
Isoniazid
Isoretinoin
Lead
Lysol
Mefenamic acid
Melarsonyl
Mepacrine
Mepazine
Mephalan
Mephenytoin
Meprobamate
Mercaptopurine
Mesoridazine
Methacycline
Methapyrilene
Methimazole
Methotrexate
Methylclothiazide
Methyldopa
Methylene blue
Methylphenobarbitone
Mitomycin
Mustard gas
Nalidixic acid
Napthalene
Neomycin

Niridazole
Nitrites
Nitrobenzene
Nitrofurantoin
Nitrofurazone
Novobiocin
Oral contraceptives
    (some)
Oxacillin
Oxyphenbutazone
Pamaquine
Penicillins (some)
Pentamidine
Pentaquine
Phenobarbitone
Phenothiazines
Phenylbutazone
Phenylhydrazine
Phenytoin
Phytonadione
Pipamazine
Piperazine
Plasmoquine
Prilocaine
Primaquine
Primidone
Probenicid
Procainamide
Procarbazine
Propylthiouracil
Pyrazolones
Quinacrine
Quinidine
Quinine

Rifamicin
Stibophen
Streptomycin
Sulfacetamide
Sulfadiazine
Sulfamethizole
Sulfamethoxazole
Sulfamethoxypyridine
Sulfanilide
Sulfasalazine
Sulfisoxazole
Sulfonamides
Sulfones
Sulphonylurea
Suramin
Tetracycline
Thiazolsulfone
Thiocyanate
Thiosemicarbazones
Tolazamide
Tolazoline
Tolbutamide
Triamterene
Trichlomethiazides
Trifluoperazine
Trimethadione
Trimethoprim
Trinitrotoluene
Tripelennamine
Tyrothricin
Urethane
Vinblastine
Vitamin K analogs

COMPOUNDS CAUSING REDUCTIONS
OF NEUTROPHIL NUMBERS

Acetazolamide
Acetoaminophen
Acycolvir
Allopurinol
Aminoglutethimide
Aminosalicylic acid
Amitriptyline
Amodiaquine
Amoxapine
Amoxicillin
Antimony
Antipyrine
Aprinidine
Azathioprine
Captopril
Carbamazepine
Carbimazole
Cephalosporins (some)
Chlorambicil
Chloramphenicol
Chlordiazepoxide
Chloroquine
Chlorpromazine
Chlorpropamide
Cimetidine
Ciprofloxacin
Clindamycin
Clozapine
Colchicine
Co-trimoaxlzole
Dapsone
Desimpramine
Diazepam
Ethacrynic acid

Ethanol
Ethoximide
Etopside
Fenoprofen
Flucytosine
Fluodarabine
Fluphenazine
Flutamide
Furosemide
Gentamicin
Gold salts
Griseofulvin
Hydralazine
Hydroxychloroquine
Ibuprofen
Imipramine
Indomethacin
Interferon alpha
Isoniazid
Levamisole
Levodopa
Lincomycin
Mepazine
Meprobamate
Methacycline
Methazolamide
Methicillin
Methyldopa
Methylpromazine
Methylthioural
Metronidazole
Miaserin
Minocycline
Oxacillin
Oxyphenbutazone

Penicilliamine
Penicillins (some)
Perchlorate
Perphaenazine
Phenidione
Phenothiazine
Phenylbutazone
Phenytoin
Procainamide
Promazine
Propranolol
Propylthiouracil
Pyrimethamine
Tamoxifen
Tetracycline
Thiazide diuretics
Quinidine
Quinine
Ranitidine
Rifampin
Spironolactone
Streptomycin
Sufasalizine
Sulfonamides (some)
Thioacetzone
Thiocolchicine
Tolbutamide
Trimethadione
Trimethoprim
Trimetrexate
Valproic acid
Vancomycin
Vitamin A
Warfarin
Zidovudine

COMPOUNDS AFFECTING NUMBERS, FUNCTION, AND AGGREGATION OF PLATELETS

Antihistamines
Aminopyrine
Ampicillin
Caffeine
Calcium channel
  blockers
Cephalosporins
Cyclooxygenase
  inhibitors
Daunorubicin
Dextran
Dipyridamole
Ethyl biscoumacetate
Fibrinolytic agents
Frusemide
Heparin
Hydroxyethyl starch

Ibufenac
Indomethacin
Ketoprofen
Mefenamic acid
Methicillin
Mithramycin
Nifedipine
Nitrofurantoin
Nitroglycerin
Nitroprusside
Nonsteroidal anti-
  inflammatory drugs
Pencillins
Phenylbutazone
Phosphodiesterase
  inhibitors
Prooxyphene

Propranolol
Prostaglandin E_2
Prostaglandin I_2
Pyrimethamine
Quinidine
Streptokinase
Theophylline
Thromboxane
  sythase inhibitors
Ticlodipine
Tissue plasminogen
  activator
Tricyclic
  antidepressants
Urokinase
Warfarin

Appendix **G**: Leukocyte—Cluster of Differentiation

The cell surface markers of leukocytes can be identified by using monoclonal antibodies that react with the different antigens on the cell surface. These antibodies have been given a designated common cluster of differentiation (CD) number. The letters *a* and *b* are used for antigens that are very similar, and the letter *w* indicates a provisional CD number assignment. The original thoughts that these antigen CDs were specific to some cell lineages have been shown not to be correct in several cases, with some of the surface antigens being expressed by cells of different lineages and some markers being more useful when measuring and detecting the hemopoietic progenitor cells. Although the number of CD markers now totals more than 200 for human markers, the markers available for laboratory animals are less numerous (see references in Chapter 7).

Most suppliers of these monoclonal antibodies provide up-to-date information on the suitability and cross-reactivities of these antibodies. Many antibodies are conjugated to fluorochromes (e.g., fluorescein isothiocyanate [FITC] and phycoerythin [PE]) to enable cell detection in either blood by flow cytometry or tissues, but again, the available combinations of conjugated antibodies can prohibit the application of double and triple labeling techniques with samples from laboratory animals. The specificity and avidity of antisera vary with supplier and sometimes with antisera batches from the same supplier. In the following table, some cell types are italicized to indicate the possible weaker binding with an antibody. To eliminate nonspecific fluorescent events, it is important to use a negative control that has a similar isotype to the antisera being employed.

A number of international workshops on human leukocyte cluster differentiation antigens have been held, and new CDs continue to be assigned.

See Zola, H., and Swart, B., "The Human Leucocyte Differentiation Antigens (HLDA) Workshops: The Evolving Role of Antibodies in Research Diagnosis and Therapy," *Cell Research* 15 (2005): 691–94.

CD Number and Cellular Expression

CD1	Cortical thymocytes, Langerhans cells, *B cell subset dendritic cells*
CD1b	Cortical thymocytes, Langerhans cells, *B cell subset dendritic cells*
CD1c	Cortical thymocytes, Langerhans cells, *B cell subset dendritic cells*
CD2	T cells, *thymocytes, NK cells*
CD2R	Activated T cells, NK cells
CD3	Thymocytes, mature T cells, T cell receptor antigens
CD4	T helper/inducer cells, monocytes, macrophages
CD5	Thymocytes, T cells, B cell subset
CD6	Thymocytes, T cell subset, B cell subset
CD7	Majority of T cells
CD8	T cytotoxic/suppressor cells, NK cells
CD9	Pre-B cells, monocytes, platelets
CD10	Lymphoid progenitor cells
CD11a	Leukocytes
CD11b	Granulocytes, monocytes, NK cells
CD11c	Granulocytes, monocytes, NK cells, B and T cell subsets
CDw12	Granulocytes, monocytes
CD13	Myeloid monocytes, granulocytes
CD14	Monocytes, some granulocytes and macrophages
CD15	Granulocytes
CD16	NK cells, granulocytes, macrophages
CDw17	Granulocytes, monocytes, platelets
CD18	Leukocytes
CD19	Pan B cell
CD20	Pan B cell
CD21	Mature B cells, follicular dendritic cells
CD22	Mature B cells, follicular dendritic cells
CD23	Activated T cells, B cells, macrophages, eosinophils, platelets
CD24	B cells, granulocytes
CD25	Activated T cells, B cells, macrophages
CD26	Activated T cells, B cells, macrophages
CD27	Thymocyte subset, mature T cells, NK cells, platelets
CD28	T cell subset
CDw29	CD4 subset, B cells, monocytes/macrophages
CD30	Activated T and B cells (Reed-Sternberg cells)
CD31	Platelets, monocytes, macrophages, granulocytes, B cells
CDw32	Monocytes, granulocytes, B cells, eosinophils
CD33	Myeloid progenitor cells, monocytes
CD34	Hematopoietic precursor cells, capillary endothelial cells

CD Number and Cellular Expression (continued)

CD35	Granulocytes (basophils' negative), monocytes, B cells, erythrocytes, some NK, cells
CD36	Monocytes, macrophages, platelets, B cells weakly
CD37	Mature B cells, T cells, and myeloid cells
CD38	Plasma cells, thymocytes, activated T cells
CD39	Mature B cells, monocytes, some macrophages, vascular endothelium
CD40	B cells, monocytes weakly, carcinoma cells
CD41	Platelets, megakaryocytes
CD42	Platelets, megakaryocytes
CD43	Leukocytes, not peripheral B cells
CD44	Leukocytes, erythrocytes, platelets (weakly), brain cells
CD45	Pan leukocyte
CD46	Hemopoietic and nonhemopoietic cells (not erythrocytes)
CD47	All cell types
CD48	Leukocytes
CDw49b	Platelets, activated T cells
CDw49d	Monocytes, T cells, B cells, thymocytes
CDw49f	Platelets, megakaryocytes, T cells (weak)
CDw50	Leukocytes
CD51	Platelets
CDw52	Leukocytes
CD53	Pan leukocytes
CD54	Endothelial cells, many activated cell types
CD55	Many hemopoietic and nonhemopoietic cells
CD56	NK cells
CD57	NK cells, T cells, B cell subsets
CD58	Many hemopoietic and nonhemopoietic cells
CD59	Many hemopoietic and nonhemopoietic cells
CDw60	Platelets, T cell subset
CD61	Platelets, megakaryocytes
CD62	Activated platelets, endothelial cells
CD63	Activated platelets, monocytes, macrophages
CD64	Monocytes
CDw65	Granulocytes, monocytes
CD66	Granulocytes
CD67	Granulocytes
CD68	Monocytes, macrophages
CD69	Activated T and B cells activated macrophages, NK cells
CDw70	Activated T and B cells, (Reed–Sternberg cells), macrophages (weakly)
CD71	Activated T and B cells, macrophages, proliferating cells

CD Number and Cellular Expression (continued)

CD72	Pan B cells
CD73	B and T cell subsets
CD74	B cells, macrophages, monocytes
CD75	B cell subset
CD77	Activated B cells, follicular center B cells, endothelial cells
CD78	B cells
CD158 to CD160	NK cells
CD233 to CD242	Erythrocytes

Index